The Image of Eternity

The Image of Eternity

Roots of Time in the Physical World

David Park

A MERIDIAN BOOK

NEW AMERICAN LIBRARY

TIMES MIRROR

NEW YORK AND SCARBOROUGH, ONTARIO
THE NEW ENGLISH LIBRARY LIMITED, LONDON

NAL BOOKS ARE AVAILABLE AT QUANTITY DISCOUNTS
WHEN USED TO PROMOTE PRODUCTS OR SERVICES, FOR
INFORMATION PLEASE WRITE TO PREMIUM MARKETING
DIVISION, THE NEW AMERICAN LIBRARY, INC.,
1633 BROADWAY, NEW YORK, NEW YORK 10019.

Library of Congress Cataloging in Publication Data

Park, David Allen, 1919-
    The image of eternity.

    Reprint of the ed. published by University of
Massachusetts Press, Amherst.
    Bibliography: p. 143
    Includes index.
    1. Time   2. Space and time.   3. Motion.   4. Physics
—Philosophy.   I. Title.
[BD638.P37   1981]   115   81-289
ISBN 0-452-00551-5   AACR1

This is an authorized reprint of a hardcover edition
published by the University of Massachusetts Press.

SIGNET, SIGNET CLASSICS, MENTOR, PLUME,
MERIDIAN and NAL BOOKS are published *in the
United States* by The New American Library, Inc.,
1633 Broadway, New York, New York 10019, *in Canada*
by The New American Library of Canada Limited,
81 Mack Avenue, Scarborough, Ontario M1L 1M8, *in the
United Kingdom* by The New English Library Limited,
Barnard's Inn, Holborn, London EC1N 2JR, England

First Meridian Printing, April 1981

1 2 3 4 5 6 7 8 9

PRINTED IN THE UNITED STATES OF AMERICA

To Edwin Avery Park

Over the years, our conversations

touched all these questions.

Contents

Preface

Time is reckoned a tricky subject, elusive precisely because its nature is to elude. We live in an instant: the instant is gone. What is it that has gone? Where did it go? The experience of time is hard to discuss in rational language; we use metaphors to deal with it and end by taking them for literal fact. Yet natural scientists take account of time in an objective fashion. They measure the time of physical processes and represent these measurements by symbol and number in equations that state the laws of the universe. Under the even light of reason it is difficult to see the shadows that troubled us a moment ago.

This book was written in the belief that science and metaphors are talking about the same things in different terms and that each can help explain the other. Part of the pleasure in writing it has been the discovery that even before there was any science in the modern sense, thinkers of the past were contributing to the synthesis that is possible today.

My main concern has been to show how the temporal flow of the universe runs through us all. The old people expressed this thought in myth. Modern cosmology, an imaginative construction which itself perhaps deserves the name of myth, furnishes another way to say the same thing. We are linked with the cosmos, body and mind, we are made of its substance and obey its laws, yet the universe that is an object of our understanding is, in a sense that I will try to explain, the creation of human minds. We describe the universe within the same context as we describe our own actions when we contemplate them objectively: extended in time—past, present, and future. But there is another mode in which we experience time when we live it, when our minds are conscious only of the present. Why the way we analyze and explain time is so different from the way we experience it is, I think, a proper scientific question, but it is one that remains unanswered. This question hovered over some of the earliest philosophers, though they identified it and dealt with it in characteristically different ways. And it persists even today as a puzzle in physical theory.

A word about texts. For the Presocratics I have generally followed the interpretations of Kirk and Raven 1957; for Aristotle, I have used the Oxford University translation edited by W. D. Ross and conveniently abridged in Aristotle 1941. Since precision and clarity of expression are so often complementary virtues, I have sometimes, after consulting other translations and the original, partly rephrased them in more colloquial language. Other translations are explained or attributed in the footnotes.

Through the years I have included some of the material in this book in a course on the natural philosophy of time, taught at Williams College. I have learned much from many thoughtful research papers my students have written, whose contents have been absorbed into lecture notes and now into this book.

I must thank J. T. Fraser and Denis Corish for thoughtful and erudite readings of an earlier version of this manuscript. My wife, Clara Park, has explained Greek texts to me and has, as always, done nothing to impede my work.

David Park

Introduction:
What Is There to Talk About?

1

I could tell you what an alternating-gradient synchrotron is and you would probably believe me.

I could distinguish between Thrones, Dominations, and Powers in the hierarchy of Heaven and you would follow my meaning, although you might doubt that such distinctions are really necessary.

But I cannot tell you what time is, because you already know. St. Augustine (the African one) began a brilliant analysis with the complaint, "What, then, is time? If no one asks me, I know what it is. If I wish to explain it to him who asks me, I do not know." [1] But really he did know, and the trouble is precisely that his interlocutor knows also, and the two knowledges are not the same. For one person time is a scale along which events are measured. For another it is events themselves in their ceaseless flow. For another it is the sum of hope, fear, doom, and joy, all in turbulent motion. How can ideas so clear or at least so vivid, but so different be brought into harmony with each other? Whatever your image of time may be, it will not be changed by anything I say. All that can be hoped is that in reading this book you will come to grasp a little more of the richness of concepts of time, and to understand how other people have thought about it. With great respect for the poets, philosophers, and historians who have written on other aspects of time, I shall scratch around in my own garden, which is the study of physics, and try to make bloom some of the questions and insights which this particular scientific method has to offer.

Physics is as much a trait of mind as it is a body of knowledge. It is the imperative, Simplify! This may sound the purest hypocrisy in the ears of those who have wandered through the observation halls of a great accelerator or stared at a mathematical formula two pages long. But that is the problem of getting to understand the subject, to understand the impulse that carries a physicist through thickets of wires and forests of symbols toward the imagined ultimate simplicity in which we really understand with our inner intuitive sense how the universe of atoms, stars, and galaxies fits together. And on a smaller scale the aim of this book is the same: to show how by asking the right questions and ignoring the wrong answers we can arrive at a simple and definite physical basis for the study of time, a vocabulary and a point of view and a few facts that make time—as it enters our description of nature and of ourselves—no more mysterious than space. But who says that space is not mysterious? Or existence? It is human consciousness we should stand in awe of. If you cherish mystery, be assured that that enigma at the center of philosophy remains before us intact as it has for three thousand years. If you favor clarity, I shall try to put things in order so as to show some system in our conceptions of the world, conceptions among which time plays a vital role.

The scientific culture is of course only one of the cultures that exist on the globe. But physical science, which is part of our own culture's system of ideas, provides penetrating and specific insights around which we can organize some, at least, of our concepts of time. The statements of science, unlike those of any other system of thought, come with recognized standards by which they may be accepted or rejected, and most rational people can agree on how these standards should be applied. But note that I do not claim we can know certain theories are true. I think the word "true" should be saved for situations like those in mathematics where a valid proof exists and all argument comes to an end; we shall never have such truth in any natural science. Though scientific theories grow out of a particular culture, they seem to be in themselves remarkably free of cultural bias, for we find them being competently judged and used by people from many countries. This universality encourages the idea that scientific insights about time represent relatively fixed points in the swirling tides of human intuition and form a natural starting point for an inquiry into what we are talking about when we try to discuss time in a relatively impersonal way.

Even under this aspect, time offers difficulties and subtleties. Things would be simpler if we could stay within the restricted context of experience on earth, our laboratory. But we cannot. The entire universe must come under our gaze, and we must situate ourselves with respect to its scale and history. Though we might prefer to discuss questions of time apart from questions of space, Einstein has shown us that the two are mathematically inseparable. So we shall have to work into the subject from first principles and determine what it is necessary to know. Out of the discussion will finally emerge some answers to questions encountered while exploring the wider shores of time.

Can time be brought into our descriptions of nature in a simple way?

Our knowledge of the past, and even more so of the future, seems to be of a different kind from our knowledge of the present; what is the relation between these different kinds of knowledge?

How are human experiences of time related to the evolution of the universe as a whole?

Does time really pass?

How is time related to space?

What is the human meaning of scientific time?

Very pretentious questions, especially the last one, and yet even in the incomplete state of our knowledge and the weakness of our reasoning powers, something can be said about each of them.[2] "More brain, O Lord," groaned Valéry, and it would indeed be nice. But for the present we must struggle along with what is available and use it the best we can. It will be well to start with the ancients, not only for historical reasons but also because in the earliest phases of a discussion, before opposing strategies and points of view have been worked out, it is sometimes possible to see with special clarity where difficulties lie.

A little reflection shows that the difficulties lie largely in the area of language and logic. After all, we are perfectly familiar with the wordless concepts that belong to time, and we easily manage the physical and social activities that involve them. As Augustine complained, the difficulty starts when we try to explain what has been going on. When we do, we become involved in overlapping problems which can be separated into problems of fact, of logic, and of strategies of explanation. I will deal with logical difficulties first, because the discussion will be rather short and because it will illuminate questions of strategy. Facts will accumulate as they were learned, in historical sequence.

Logic

The old Greeks regarded themselves as late arrivals on the intellectual scene. Plato described how the Egyptian priests plied Solon with their ancient civilization until he felt like a schoolboy from a country town, and even the great geometrical truth of Pythagoras (that the lengths of the sides of a right triangle satisfy the relation $a^2 + b^2 = c^2$) was known to the Babylonians a thousand years earlier.[3] But when the Greeks arrived they produced, almost at once, an entirely new idea which was destined to have great consequences: Pythagoras, or at any rate someone in the seventh century B.C., invented the idea of mathematical proof. That is, if one grants certain hypotheses, the proof shows that one must grant the conclusion. We do not know how the Babylonians regarded the rule of right triangles, whether as a convenient recipe, as a law of nature, or even perhaps as a theorem previously proved, but Pythagoras and his followers left no doubt as to how they regarded it: it was Truth.

Succeeding generations of mathematicians rapidly built up a large corpus of proven mathematical knowledge. Inevitably, people began to wonder whether similar methods would enable them to establish truth, once for all, in cosmology, politics, and human right and wrong. Thus logic was born, and readers of Plato and Aristotle and the Medievals know the inconclusive results of logical argument applied to these subjects. To be effective at all, logic must be embedded in a matrix of unanalyzed and unstated assumptions, and this is how it is normally used. Isolated from these assumptions it helps us understand the world and ourselves about as much as does the knowledge that $2 + 2 = 4$.

As an example, there is a serious difficulty that arises when we try to apply even simple logical standards to questions involving the passage of time. Suppose I say, "Alice is in the living room." If it is clear who Alice is and what room I am talking about, there are twenty ways of verifying my statement, and we assume (on the basis of experience) that all rational people can find out and agree, one way or another, whether or not the statement is true. But if I say, "Alice *was* in the living room," the situation is quite different, since the means of verifying the statement are different and not so sure. Traces of Alice could have been deceptively left by someone else; memory tricks us; and if I say she was there at 2 A.M. on October 3, 1955, people will shrug their shoulders. How much more doubtful if I say that Alice *will be* in the living room. Law courts are obliged to accept evidence that makes it very probable that she *was* in the living room, even though the standards of verification may not be as high as those we apply if told she *is* there. But the word "evidence" is never used when we are supporting a prediction. We feel it is overwhelmingly probable that the sun will rise tomorrow, yet we cannot claim there is solid evidence that it will. What is the basis of our certainty? And what do we do about time in logical discourse? These two questions have excited an immense amount of philosophic thought. As to time in logical discourse, I can only sample a little of what has been said. As to tomorrow's sunrise,[4] and other more or less certain cases of prediction and its opposite—which is generally called retrodiction—we shall see in chapter 4 exactly how that works. In fact the two questions are not so separate as perhaps they first appear, and my remarks about prediction may help clarify the situation in logic.

Aristotle's Puzzle

One of the cornerstones of logic is the law of the excluded middle: every unambiguous statement is either true or false. But, says Aristotle, consider the statement, "There will be a sea battle tomorrow."[5] The law states that either this is true or it is false now, in the sense that if we wait till tomorrow we will *know* it to have been true or false at the time it was made. "Now if this is so, nothing is or happens by chance, now or in the future, and there are no real alternatives. Everything that happens has to happen, in order that whoever has predicted it shall be either right or wrong." This is of course insupportable. It may be that our role in the world is that of automata, unable to decide for ourselves whether to fight or flee, but the fact, if it is a fact, cannot be established by this kind of argument. Either (*a*) the law of the excluded middle must be refined, or (*b*) one must introduce qualifications and refinements in the simple statement "This is true," distinguishing "true" from "necessarily true." It is not easy to follow Aristotle's subsequent discussion. Medieval commentators tended to think he favored (*b*); the moderns, more at home with new logics that generalize Aristotle's, think he leaned toward (*a*). In the course of this book I shall slowly transform the discussion a little, but nothing I say will make it easier.

Logical inferences that do not involve time in any way may be relatively easy to make and to justify, but as soon as processes in time come into question, the simplicity of our reasoning and the certainty of our knowledge begin to deteriorate. To be able to explain in words some of the certainty that is apparent in our wordless dealings with time, it will be necessary to deal with the question, What is the relation between the present and other times? There is a trivial answer: draw a long line, label one end "past," the other end "future," and add a dot in the middle to represent "present." This is a picture of time. There are, to be sure, modes of thought for which this picture is perfectly adequate, but Aristotle's example, and our own experience of a present perceived and conceived in a special way, suggest that there is more to it than this. The line is relatively easy to talk about, and is the main subject of this book. The point signifying "present" eludes the methods of physics, and I shall have to talk about it unscientifically (and much more briefly). The relation between the point and the line is an old puzzle, and it will be interesting to see how it arose long ago in Western thought. I do not think it can be understood at all until we understand the harmony between mind and nature that allows us to be so at ease in arranging our lives in accordance with temporal relations that we cannot really explain.

The Ancient Obsession

2

In the archaeological collections of Europe are an endless number of flint and bone tools attributed to the various ages of prehistoric man. These collections house a few carved works of art, some of which show an almost uncanny artistic vision, an ability to see the forms that underlie the surface variety of humans and animals. There are also in these collections some strange fragments which seem to have led a double life. Typical are some bone knives and scrapers that became worn or damaged and were then inscribed with a series of small marks made with a sharp stone point. In the 1960s Alexander Marshack, an American writer, became interested in these objects, and it occurred to him to have a careful look at them. Figure 1 shows one object he examined, a scraper from early Cro-Magnon times, about 30,000 B.C., bearing a serpentine arrangement of scratches. Below the scraper is a drawing of these scratches as seen through a low-power microscope. The drawing shows the variety of different strokes employed and suggests that they were made in a linear sequence.

There are several dozen such objects, taken from sites as diverse as Czechoslovakia, the Ukraine, western France, and the Congo. Their dates span at least 20,000 years. All may not have been made for the same reason, whatever that was, but they betray a common concern with sequence, with order, probably with counting. On many of them, Marshack claims to have found sequences of marks that correspond to what would result if one recorded diagrammatically the phases of the moon, night after night. But this is put forward only as a hypothesis, and even if one studies the marks carefully, it is possible not to be convinced. Still, the marks denote sequences of some sort, and the fact that they were

1a. Bone scraper, Aurignacian, from Abri Blanchard, Dordogne, France. Length 4¼ inches, date ca. 32,000 B.C.
1b. Microscope sketch showing the sixty-nine marks and distinguishing the strokes by which they were made. The gap in the center contains traces of three other marks. By permission of A. Marshack and McGraw-Hill Book Co.

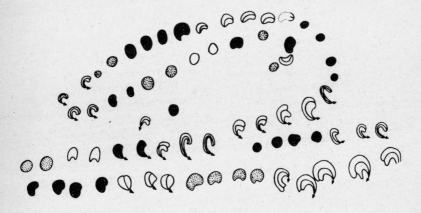

made with different stone points suggests that they were not all made in an afternoon. It is hard not to conclude that they are records.

The carvings and these strange sequences afford only the smallest glimpse of what went on in the Stone-Age mind. We do not even know, and probably will never know, about Stone-Age language, though there is evidence that words date from the last glacial period, about 50,000 B.C. (Harnard, Steklis, and Lancaster 1977). But at least we know that our ancient ancestors had two interests that distinguished them from animals: form, which we see in their carvings and paintings; and quantity.

Megalithic "Observatories"

Thousands of people annually visit Stonehenge a much-damaged and much-restored arrangement of huge stones on Salisbury Plain west of London. Substantially fewer have stopped to explore other smaller structures, generally circular in form, that cover Great Britain, Ireland, and the surrounding islands. There are more than 900 of these, and though it is impossible to date a piece of weathered stone, fragments of pots and tools found with them show that most were constructed between 3000 and 1500 B.C. Stonehenge was constructed, in several successive stages, during this period. It used to be regarded as a place of worship in the charge of the priestly cult of Druids, who flourished in England for a thousand years starting about 500 B.C. (and who now, on certain days, put on their robes and flourish again). But radiocarbon dating, with its modern corrections, places the first of Stonehenge's three stages of construction close to 3000 B.C., so it has nothing to do with Druids at all, but puts it back in the late Stone Age, among folk who until recently were considered, intellectually speaking, as howling savages.

There seems to be a good deal of art and science in the design of Stonehenge. Stones weighing many tons have been set in place with tolerances of a few inches. The general pattern of the largest stones is called a "dolmen" on the Continent: two verticals supporting a horizontal. But at Stonehenge the rough Continental form is purified and refined by precise shaping and alignment to produce, even in its ruined state, an impression of awesome grandeur.

The worst mistake we can make is to try prematurely to explain why Stonehenge was built. To use words like "temple" or "observatory" is to impose our own purposes and categories on people profoundly different from ourselves. It does not help to study the reports about stone-age people now, or recently, inhabiting the earth, since these are precisely the people who never burned with the energy that drove the remote European and Mediterranean ancestors of most of us to finish with stone and go on to the techniques and social forms that have led us where we are today. Nothing could be more closed to us than the minds of the builders of Stonehenge. We do not know what they did, or thought they were doing, or why, but there is one thing we can do: state clearly what astronomical observations can be made, with our own knowledge, using only the stones they left behind.

The central axis of Stonehenge is aligned with respect to the point at which the sun rises on Midsummer's Day, or rather, at which it rose about 5,000 years ago, since the celestial angles have changed a little. There may have been a pair of stones across which you could sight to get the exact direction, but one—presumably near the center of the circles—is gone, and the other has tilted far out of the vertical. If the English climate played one of its tricks and the sunrise could not be seen on that day, smaller stones were placed so that Midsummer's Day could be determined from observations a few days earlier and later.

It is easy to object, What is the point of being able to do this? If they were that interested, they must have known from their observations that there are 365 1/4 days in a year, and so they could find by counting, without having to get up early in the morning, when Midsummer's Day was going to occur. A possible answer is that the length of the year was not 365 days and 6 hours but about 365 days, 5 hours, and 49 minutes; that having (we believe) no clocks, they knew nothing of hours and minutes and could count only whole days; and that they were presumably in the process, slowly, of discovering that you could keep good track of the calendar by counting 365 days per year and inserting 8 extra days, distributed as evenly as possible, every 33 years.

Two astronomers, Gerald Hawkins and Fred Hoyle, have analyzed very carefully the directional alignments obtainable by sighting across various pairs of stones in Stonehenge. They have found that a few significant directions occur again and again: to the point where the sun breaks the horizon furthest north, on Midsummer's Day; to its southernmost point, reached 6 months later; and to the equinoctial point, reached in between. (If you miss the sunrise you can catch the sunset by sighting in the opposite direction.) Furthermore, in its path through the sky, the moon follows the sun roughly but not exactly, oscillating to the north and south of the sun through an angle a little greater than 5.1 degrees. These directions are also present in the Stonehenge alignments, given by auxiliary sightings that are 5.1 or 5.2 degrees north and south of each of the principal solar directions. (None of these directions is correct now; they must all be calculated back to the date when the stones were erected.)

What was the lunar information for? We must guard against assumptions implicit in the question, but it could have been used for predicting eclipses, both of the sun, which are caused by the moon, and of the moon. A lunar eclipse is caused by the earth's shadow falling on the moon, and is visible to anyone who can see the moon at the time, i.e., about half the people in the world. Thus about half the lunar eclipses predictable at Stonehenge would actually have been seen there. Solar eclipses are total for those people living within the borders of a long strip a few hundred miles wide where the moon's shadow passes, and they are partial in a broader band. Thus the number of successes would have been smaller, though each would have been impressive.

Stonehenge, its record garbled by the careless reerection of fallen stones and by centuries of vandalism, is actually poor evidence on which to rest a case for Stone-Age astronomers. There are hundreds of other stone circles in the British Isles, extending to the Orkneys and even the Outer Hebrides, and it is possible to interpret at least a few of these, too, as solar and lunar observatories. In the last few years, while road building, agriculture, industry, and plain vandalism have been destroying these monuments at a terrifying rate, they have come under increasing study. Yet it is very difficult. Those that have survived tend to be in the most inaccessible places and are often ruinous and incomplete, with some stones entirely buried in the ground and others carried away to patch up a wall. To obtain measurements precise enough to analyze critically their possible use requires surveying under hardship conditions to accuracy of inches. And to this surveyor's skill the investigator must add those of the archaeologist and positional astronomer. The most assiduous student of these monuments is Alexander Thom, Emeritus Professor of Engineering Science at Oxford, who since the mid-1950s has been finding them, measuring them, and explaining them as observatories (Thom 1967, 1971, 1979). Archaeologists vary in their reactions. It seems that some of the builders were capable of accurate surveying, that some of them used a common unit of measure—about 2.72 feet, which now is called a "mesolithic yard" (though they probably knew it by some other name), and that some alignments, often extending to distant horizons, seem closely related to the comings and goings of the sun and moon. Apparently stars were of little or no interest. A few years may clear up the mysteries, but archaeoastronomy, as it is now called, has become a legitimate field of science.[2] And even if the remains in Britain and Brittany, which are the earliest and most extensive known, are finally explained, there remain hints of similar questions about the monuments of Egypt and of Central and North America.

The Road of Excess

It is hard to imagine what our civilization would have been like without the continued expenditure of quixotic effort. There have always been projects that seem out of proportion to necessity. You do not need a pyramid to mark the remains of a king, and besides, there seem to be so many more pyramids than there were kings that many may have been erected just so there would be another pyramid (Mendelssohn 1974). The Crusades were launched, one after another, with nothing substantial to be gained from any of them. The plays of Shakespeare, the Imperial Court at Peking, the walk on the moon—all illustrate a quality of disproportion so common in enterprises with which we are familiar that we tend hardly to notice it. There are human societies that have lived peacefully, scarcely varying from one century to the next, in a tradition that matched effort to need, but even these societies may ignite without warning. A few generations before the pyramids were built, the Egyptians were living in the millennial style of peasants and fishermen along the Nile. Later, a single life—Mohammed's—was enough to generate among the camel-trading tribes of Arabia one of the greatest movements for disseminating force and culture that the world has seen.

In judging the British stones or the elaborate Mayan calendar (Coe 1971), as efforts disproportionate to the need we must be careful because we probably do not understand the need. Certainly it did not lie in the realm of practicality, and if we finally speak of effort for its own sake, for the pure pleasure of conquering abstract difficulties, we may not be far wrong.

The point of these remarks is that the rhythms of the universe seem to have held immense interest for many of our ancestors. One should be careful of loose talk, but it is as if they felt a great need to harmonize the rhythms of the universe with the rhythms of human life. In his deeply intuitive reconstructions of the primitive mind, Mircea Eliade has shown how this need has manifested itself in myth and ritual;[3] the remains we have discussed embody it in mathematics and in stone. The works to be discussed next embody it in logic and in words.

The Greek Questions

3

The ancient world consisted of cultures based, like those of today, on applied intellect. The reasoning and the experiments that led to the development of the first strong metals and alloys are lost to us, but the tools and ornaments still exist. The Ashoka pillar outside Delhi is made of iron and has withstood over 1,500 years without rusting. Indian metallurgists must have been experimenting much earlier in order to produce that huge bar of metal. The Great Pyramid is aligned accurately to within a few inches along the cardinal directions of the compass, and part of even that small error seems to be the result of Africa's slow rotation by the continental drift. Let these examples stand for many. The old people had wide and varied knowledge. Had they theories, or was it rule-of-thumb? We might hope to find some day a myth with the clearly understandable content of a scientific theory. Some claims have been made, but there is as yet no totally convincing example.[1] Even though many myths clearly *describe* agricultural or astronomical knowledge, none seeks to explain it scientifically. Whatever words were uttered perished with the speakers, even in the cultures of Egypt and Mesopotamia where there was writing, and acres of stone and clay were squandered to render mundane transactions immortal.

Anaximander

As Greek culture spread through the eastern Mediterranean it announced itself with a buzz of speculation. Every region had its thinkers and writers, and we know them by name and reputation even when their works are scanty. They quoted each other, and the later encyclopedists quoted them all. Although we might say today that they sought the laws that govern the stars, planets, and weather, as well as the more hidden laws of society and human personality, it would be more accurate to say that they were seeking the concept of law itself, and the concept was long in emerging in anything like its present form. The very word "law," used to refer to nature, is the same as the word for statute or decree in Greek as well as in English.[2] This suggests questions we have learned to treat as irrelevant: Who made the law? Why was it made? How is it enforced? The earliest philosophic text we have, widely quoted in antiquity and almost certainly authentic, is a precious fragment on law from Anaximander. Born in about 610 B.C. in Miletus, on what is now the west coast of Turkey, Anaximander lived a century before the Persian War, and his great-great-grandson might have seen the new Parthenon. He imagined the world as immersed in an endless, inexhaustible reservoir of what we today would separate out as matter, energy, time, space, and design. But to Anaximander, distinguishing no such categories, it was simply "The Boundless." Three centuries later, Aristotle summarized:

> As a beginning, it must also be something that has not begun and cannot pass away. For whatever has begun must necessarily end, and all passing-away likewise has an end. It is itself without beginning, but is rather thought to be the beginning of everything else. It encompasses all things and governs all things . . . and this, they say, is the Divine.[3]

Anaximander describes how the world arises out of the Boundless. Then he makes his great statement (Kirk and Raven 1957, p. 107):

> It is necessary that things should pass away into that from which they are born. For things must pay one another the penalty and compensation for their injustice according to the ordinance of Time.

These are the words of the law court. To their users the just and the equal were nearly synonymous, and one can think of the rhythms of nature—the alternation of growth and decay of trees in a forest, the change of the seasons, the coming into being and passing away of ourselves and the objects around us—as belonging to a vast cosmic equity, governed by the principle, "Nothing too much."[4] Anaximander's statement has little relation to modern science, but two things about it speak to us today. The first is its confidence that nature is lawful, even though the law referred to is that of the courts, the second is its tone of the highest seriousness. The question of law is important because one must finally formulate the relation that exists between human enactments and the divine. "Everything is full of Gods," said Thales, Anaximander's older contemporary. And philosophers even down to Plato, 200 years later, cite this saying again and again as the quintessence of all philosophy. But these gods were not those named personalities in Homer and the dramatists; rather they were smaller foci of power and meaning. "The fact of success," wrote Aeschylus, "is a god," and so were stars and trees and the thrill of recognizing a friend after a long absence. The ancients were men and women like ourselves, but their categories were different.

Anaximander gives us a glimpse of natural law, expressed in archaic terms. Later historians said he was the first to establish how the sun moves through the sky at different seasons, but we know nothing of how he described these motions, and besides, we have evidence in stone that others had pursued this knowledge a thousand years earlier. The question of how to describe the lawful processes of nature, linking the past and future with the observed present, was taken up by the next generation of Greek thinkers. I shall sketch only three answers among many that were available, but all three are valid today, and the modern conception of time in physics derives from attempts to harmonize them.

The problem is to be able to describe and explain the world as we perceive it and act in it. Clearly the immense variety of sense impressions that we receive in a day does not conceal that there are regularities. The sun rose and will set; the constellations move across the night sky in orderly fashion. From farming, navigation, and cooking one derives a strong impression that there are laws binding man and nature, each a sort of contract: if we act on nature as we should, she will behave in a definite and predictable way. The problem is to state these laws clearly: Is one talking about man, nature, or the union of the two? Is there an unchanging reality underlying the world of experience? What can be said about it? Or, more simply, what is the world?

Heraclitus and Parmenides

Heraclitus and Parmenides were among the most complex and subtle of thinkers; even the paraphrases given by later historians are hard to understand. Heraclitus came from Ephesus, in Asia Minor, about 540 B.C. Parmenides was born about thirty years later in Elea, in southern Italy. Heraclitus may have written a book; we have none of it, and his reputation is based on a number of sayings, made deliberately obscure to show that there were not many people he was interested in reaching. Parmenides was known principally for a poem, written in high poetic style, in which he analyzed the mysteries inherent in the single Greek word, *esti*, "is."

I cannot summarize what we know (or opine) of works of these two philosophers. I am not competent, I disagree with many who are competent, and to argue the whole matter would take two chapters. After my introduction you may want to go to the books (Kirk and Raven 1957; Freeman 1948, 1959) and read them for yourself, but only, please, after the following preparatory exercise. Think about the question, What is the world? for a while until it makes sense and is seen as an important question. If that happens, go on to the next step and try to answer it, or at least to see what is involved in an answer. Then you will be ready to think about what is being said; to put it into your own language, realizing that the process must produce errors of transmission; to make it, if possible, your own.

Heraclitus understood very well that we will get nowhere looking at surface appearances only. "The real constitution of things generally hides itself," reads one fragment, and again: "An unapparent connection is stronger than an obvious one." Continuing in the tradition of Anaximander, Heraclitus saw conflict as the central reality in the world—conflict that strives away from a balance but, in the long run, leads back to it, just as war or litigation does. "It is necessary to know that war is common and strife is right and that all things happen by strife and by necessity." In this philosophy process is central; even things become processes. Myself today is different from myself yesterday. "We step and do not step into the same river; we are and are not." This fragment may be spurious, but it is in the Heraclitean vein, and expresses his idea as the ancients understood it.

Here we have a sketch of a solution to the problem Aristotle illustrated with a sea battle. If the sea battle is an event involving things—ships and sailors—then the question of truth in past and future is entirely different from that of truth now. But if the world is process, then what is real are the meteorological, naval, and political processes that lead toward and away from the fight. If we understand them properly then we know, as a corollary, what can be said about tomorrow's events. Time, far from being an intruder into logical discourse, becomes central to it.

Parmenides expressed himself very differently. If the nature of existence is the central question, we must decide upon the meaning of the word "exist." Parmenides arrives at it logically. He seems to have been the first to hold the idea that as proof is to mathematics, so logical argument is to philosophy, and his argument can be expressed in the following steps:

1. The object of thought is thought. (That is, insofar as the ideas in our minds are logically arrived at, they have their origin in other ideas, not in experience.)

2. What exists can be thought about.

3. Nothing can be thought about what does not exist. (Echoed in Wittgenstein's "Whereof one cannot speak, thereof one must be silent.")

4. The world exists. It is what exists. It is One.

5. The world is what it is and does not become something else.

6. Neither the world nor any part of it came into being or will pass away. (This follows from nos. 1, 3, and 5.)

7. The world is a timeless whole. (If nothing about it comes into being or passes away, it never changes.)

8. This world is not the one we perceive with our senses. (The perceived world is a world of change.)

These assertions would be easy to make into a coherent picture of the world if one wanted to talk about only timeless entities like numbers or the properties of geometrical figures. But these are only a part of what exists. What about the rest? There is, after all, a world out there, a universe. It exists, and even if we cannot know it with our senses (nos. 1 and 8) it can be thought about (no. 2). Very well, then how are we to think about it? Parmenides tells us (Kirk and Raven 1957, p. 275): the world is One.

> It is not divisible, since it is all alike; nor is there more in one place and less in another which would prevent it from sticking together nor is one part worse than any other: it is all full of what exists. So it is all continuous, for what exists clings closely to what exists.

The world is spatially finite (*Ibid.*, p. 276):

> Motionless within the limits of mighty bonds, it is without beginning or end, since coming into being and perishing have been cast out by true belief and driven away. Unchanging, it stays fast where

it is. Necessity holds it firm within the bonds that keep it back on every side, for it is not lawful that what exists should be unlimited. It does not need anything else; if it did, it would need all.

Finally, a passage (*Ibid.*, p. 273) which explains how "coming into being and perishing have been cast out by true belief":

It is entire, immovable, and without end. It was not in the past, nor shall it be, since it is now, all at once, continuous. . . .

I interpret this as meaning that for Parmenides, the idea of the world includes all its history: past, present, and future taken at once. There is no present, only the intricate web of events as eternity beholds it. In such an idea the notion of change has no place, and nothing could make clearer the distinction between what Parmenides calls the "Way of Truth" and the "Way of Seeming," which is the world as we perceive it.

This conception, tremendous in its originality and imaginative power, is one of the supreme achievements of classical thought. The time of Parmenides is that of Plato's World of Ideas, and in the next chapter I shall show that it is also the time of mathematical physics. To understand its originality, consider how Homer told a story. His narrative technique is subtle and varied; there are interpolations and flashbacks, but whatever is happening, we are in the middle of events. We are rarely asked to stand back and look at the story as a whole: Homer is to be experienced, not interpreted.[5] He reproduces the experience of life itself; Parmenides describes and explains the universe but his view is necessarily that of an outside observer. He does not narrate it.

Later commentators saw Heraclitus and Parmenides as irreconcilably opposed, but it is an opposition based on viewpoint, not on fact. For Heraclitus the world is process, and we are immersed in it. Parmenides sees the world from another dimension, as an object together with its history; he sees our birth, life, and death with a single glance. Each is right in his own way, and it is our business to understand how this can be so.

Atoms

One of the central intellectual devices for dealing with the problems of existence, permanence, and change is provided by the atomic theory, which arose in response to the questions and answers we have been discussing. In a better-ordered world it might have been invented as an answer to questions like What happens to the water in a wet shirt hung up to dry? or How does the wind exert its force when it blows? or How is it that out of the rotten remains of a dead stump a new tree grows, nourished by the remains of the old, but living and strong? But those questions, and their fairly obvious answers in terms of atoms, came later, when the theory was already clear in the minds of its adherents and they had reached the stage of explaining it to everybody else.

The original formulation of the atomic theory must have been very abstract. Of the founder of the theory, Leucippus of Miletus (or perhaps he came from Abdera), we know essentially nothing except that he was about forty years younger than Parmenides. His fragment is one weighty sentence; everything else comes from later commentators or from his more verbose pupil Democritus, who, because he is more widely known, has usurped the title of the father of atomism. Their model universe consists of atoms separated by empty space. The atoms are continually in motion. They are small and hard and come in infinitely many kinds. They have weight. "Democritus's school thinks that everything has weight but that fire, because it is relatively less heavy, is squeezed out by heavier matter. Because it moves upwards we think of it as weightless."[6] The properties of atoms reflect, in various ways, those of the substances they form. "Bitter taste is caused by small, smooth, rounded atoms whose contours are actually rippled; this makes them sticky and viscous. The taste of salt is caused by large atoms that are angular, even jagged."[7] The commentator Simplicius summarized the atomic structure of matter in words that would almost go into a modern text: "These atoms move in infinite space, separate from one another and differing in shapes, sizes, position, and arrangement; if they collide they may bounce off in any random direction or they may fit together by virtue of their shapes, sizes, positions, and arrangements to form composite objects."[8] Democritus did not distinguish between atoms and molecules, but he would not have been surprised by molecules. Atoms have a long history that begins here, but it is important to realize that this theory was purely speculative and not based on any observation of atoms;[9] in fact as late as the beginning of this century it was possible for very famous physicists and chemists to dismiss them contemptuously as a metaphor that had got out of hand. It was not until 1905 that Albert Einstein, in his first great work, showed how an experiment could be done that could not be explained unless one assumed that atoms are (in a suitably qualified sense) real.

For our purposes the great point of atomism is the one Leucippus made in his only surviving words: "Nothing happens at random, but everything for a reason and by necessity." (Kirk and Raven 1957, p. 413.) It seems an obvious conclusion if you think about it. If this is true, then the present necessarily implies the future and the past. The truth of all three is of the same kind; the sea battle will necessarily take place or not take place, and as Aristotle clearly recognized, it follows that human choices have no bearing on the matter. The Greek atomists left no escape clause. The soul and the mind are of spherical atoms like those of fire and are scattered throughout the body. The conflict between atomic theory and the agreed interpretation of the ordinary experience of life—that choice is, within limits, made freely—is absolute, and it is as real today as it was in the fifth century B.C. It requires, if not resolution, at least that the two views be put in harmony; I shall try to do this in chapter 10.

Plato

Just as the symphonies of Beethoven can be regarded as a single musical composition with structure, development, and climax on a huge scale, so the written philosophy of Plato is best understood as a single statement running through his dialogues from their beginning to their end. To consider one topic of this work is to omit the rest, and what is kept is weakened. But Plato demands to be included here because it was he who perceived that a bridge must and can be built between Parmenides and Heraclitus.

For Plato there are two kinds of what might roughly be called "knowledge." The first is what we gain from our ordinary observation of the world around us. There is nothing logically necessary about this knowledge. It is subject to all the uncertainties of sense, understanding, and recollection. Parmenides called it "seeming"; Plato calls it "opinion." There is such a thing as real knowledge, but what is known in this sense is not the world but "Ideas," the word is used in a very special way. An Idea is a prototype; nobody has ever seen a perfect circle, but everyone has a perfectly true idea of what one is, whether or not he can recite the proper definition. For Plato, in exactly the same sense, justice is an Idea. Ideas are the subjects of the verb "exist" and the objects of the verb "know," but they are more than this, for to every one pertains to some extent the highest of all Ideas, the Good.

Ideas are accessible to the prepared and disciplined mind. They cannot be logically deduced from observations of the world about us—that is the way of opinion—but they come whole, in an act of unanalyzable inspiration, perhaps out of memory, after long preparation. There are not several ideas of a circle, of justice, of the Good. The Idea is One. The word suggests Parmenides, and the formula clearly derives from him, but it goes further. Socrates treats the austere and venerable Parmenides with a respect unique in the dialogues, but completes his thought in a way that illuminates the meaning of the world and of life. Heraclitus, on the other hand, has confused sense with understanding, and opinion with truth.

Ideas are eternal in that they have nothing to do with time. This they share with the universe as Parmenides described it. The world of sense is governed by change. Where among the Ideas is the prototype of change, of time? This question poses a substantial problem for Plato's theory; we shall see in chapter 10 how it is solved.

I have not discussed the ancients for their literary value or because they are "the beginning of the story," but because they thought with great clarity and originality, unencumbered by any need to say complicated things. They realized, implicitly at first and later through close argument, that an understanding of time was essential to their task, and they sought this understanding in various directions. Finally, they came up against the question of human freedom. We shall see in later chapters that this is not the only question that must be answered before we can claim to understand ourselves as part of the world, but it is probably the hardest one. In the next chapters we shall develop the materialist side of the argument and see how, taking due account of time, the laws of nature are found, stated, and used.

The Newtonian Answer

4

Try to imagine the childhood of science. No particular scene need be assumed, no special epoch. Perhaps science was born for some Stone-Age people more than 5,000 years ago; for some moderns it has perhaps not yet started. It begins when people observe natural facts and wonder about them. The days are longer in summer than in winter. Why? The moon changes not only her shape but her path through the sky from night to night. Why? A woman's menstrual cycle occupies almost exactly one lunar month. Why? She cannot conceive a child unless she receives a few drops of fluid; the child so conceived is born after nine months and often resembles the father as well as her. Why? How does it all work? Is there any direct connection between human fertility and the moon? How could one find out? The varieties of human temperament extend from those to whom such questions are nothing at all to those for whom they are everything, and it must also have been so in antiquity.

The Motions in the Sky

Suppose your imagination has been taken by the motion of the sun. In winter it rises and sets toward the south; in summer, toward the north. If in the period between you record its setting place with a couple of stones aligned to the horizon, you discover (in northern latitudes) that its path shifts by as much as a full solar diameter from one evening to the next. Where is it during the night? Well, where are the stars during the day? Some people I have met, a little imaginative in other ways, have seemed unable to answer, and yet the answer is obvious, for after sunset the stars appear in their places in the sky. They do not rush to get there on time; no ancient myth says that. Rather, light fades as in a theater and they are revealed. They were there all the while, and comparison of the dawn and evening skies reveals that during the day they moved, invisible, overhead just as the sun did. This is the first and greatest astronomical discovery.

Is the sun, then, motionless among the stars? This can be answered if one watches sunsets during the year. No, the sun marches through a great circle of stars during a year. The ancients divided them into twelve constellations, named with regard to some cosmological theory of which only the bravest scholars think they have some knowledge.[1] Does the newspaper say you are a Gemini? According to the old ideas, that is the constellation in which the sun stands on your birthday. But do not bother to go out in the evening and verify it, for it is not true any more. The astrologers' tables were made up almost 2,000 years ago and since that time the sun has slowly fallen behind at the rate of about 1 angular degree in 72 years, or one zodiacal sign in 2,150 years. (Astrologers aware of this fact say that astrology "works better" if they ignore it.) The chances are that according to the sun you are a Taurus. But think of the shock to serious thinkers of the past when they realized that the sun does not exactly repeat his daily journey each year. Something is wrong with the universe. Will it right itself or only get worse? Somebody unauthorized got into the sun-god's chariot and drove it off its correct path. He wrecked it and died in the crash; his name was Phaethon. The myth may possibly commemorate the discovery that the circle of the zodiac is in motion with respect to the earth's equatorial plane, expressing in language everybody could understand that something terrible had happened, even if they were not quite sure what. At least, Plato interprets it this way in *Timaeus*, 22C.

The situation was particularly evident in Egypt, because each new temple built aligned to the rising of a star-god had to have a different orientation (Lockyer 1894). The discovery that the zodiac moves evenly in a motion called the precession of the equinoxes is commonly credited to Hipparchus in the second century B.C., but it is hard to believe that the Egyptians had not deduced it earlier from the architectural evidence around them. At any rate, the first step was to establish the precession by measurement. The second was its scientific explanation, and that had to wait a long time. In between there were explanations of other kinds, which we know through myth. That of Phaethon is appropriate to the seriousness with which the fact of precession was viewed but does not explain it. There could be no explanation until many more measurements had been made: of the moon and planets, of falling and rolling balls, of swinging pendulums. The scientific argument that explains precession must involve no special new hypothesis, no Phaethon; in fact, as Isaac Newton finally realized, there must be one single dynamical theory, resting on experiments but more general in that it predicts the results of others not yet performed. Only thus can order be brought into our experience of the world.

I must make clear, before explaining anything of Newton's theory, that it is its universality that distinguishes science from nonscience. It makes any science terribly vulnerable, since it can be attacked wherever it makes a statement. A single clear-cut discrepancy between theory and observation means quite simply that the theory is wrong. Scientific theories are formulated so as to emphasize this vulnerability, since when we are proved wrong we learn something. Nonescience is more careful, and protects itself by vagueness.[2] If somebody presents you with a theory of human motivation or social dynamics, ask what are the means by which the theory might be disproved. If there are none, then it is not science, and it is probably not worth much.

Isaac Newton

Isaac Newton was born in Lincolnshire on Christmas Day 1642, the posthumous child of a yeoman. Helped by people who recognized his talent, he found his way to Cambridge, where he learned much that was true and much, of course, that was not, but when the plague struck England in 1665, the universities were closed and he had to go home. Back in Lincolnshire he had time to reflect on what he had been learning, and he started to make scientific discoveries and inventions. Among the first of the inventions was a theory of the moon. He was able to calculate the length of a lunar month—the interval from full moon to full moon. The result was slightly in error, but it is important for our argument to understand what went into the calculation. Expressed in words, Newtonian dynamics reads thus: a force applied to an object changes its motion—speeding it up, slowing it down, sending it into a curve. The motion the object actually executes is the result of these changes but also, of course, depends on how it started out. In fewer words, the motion of a body can be calculated if one knows the applied force and the initial conditions. The applied force comes from outside. There may be stresses inside the object also but they do not change its motion. You cannot make a car go faster by pushing on the dashboard in front of you. The motion usually depends on the object's mass; throwing a heavy stone is different from throwing a light one. So the moon's motion depends on the force exerted on it and how the motion started.

The second ingredient in Newton's theory was an assumption about the force: that it is merely the moon's weight, pulling it down toward the earth as any massive object is pulled down, except that if something is as far away as the moon, distance becomes important and the force is reduced. There is a tacit assumption here borrowed from Galileo. Earlier philosophers had assumed that if the moon moves there must be some force to keep it going; otherwise it would stop. Galileo convinced himself that a moving object does not stop unless something stops it,[3] and Newton's hypothesis of a gravitational force on the moon was not that it functioned to keep it going but rather that it served to keep it from vanishing into the distance. Perpetually falling toward the earth, the moon perpetually misses because its motion carries it forward at the same time.

Newton found that if he made the simple assumption that gravitational attraction varies inversely as the square of the distance from the earth's center, the length of the month came out pretty well. He had also to make some assumption about the moon's initial motion. He assumed that the moon had always moved in a circular orbit around the earth. This was only an approximation. He probably realized at this time, and he certainly realized later, that there was another force: the moon gravitates also toward the sun, which though far away is very large. Much of the art of physical calculation lies in making the approximations that are appropriate to the precision you expect. Newton assumed a circular orbit and ignored the sun. The error in the result was probably a fraction of a day (we do not know exactly what numbers he used). He told nobody what he had done.[4]

The *Principia*

In 1687, twenty-one years after these early thoughts, Newton published his greatest work, the *Principia*, whose complete title in English is *The Mathematical Principles of Natural Philosophy*. It is very hard reading, largely because in making his results public Newton had used the cumbersome language of Euclidean geometry to express the dynamical results that he seems to have obtained much more simply in private by the new methods of differential and integral calculus. The book is in three parts: the first shows how to calculate the motion of an object if you know the forces acting on it; the second considers the various forces actually encountered in nature (omitting electricity and magnetism, which had scarcely been studied at the time); the third analyzes the motion of planets, moons, and comets under the assumption that the only force coming significantly into play is the force of gravity. This is the same force he had earlier invoked to explain the moon's motion, except that now the idea is vastly generalized into the assumption that a gravitational force exists between any two objects in the universe, proportional to the mass of each object and inversely proportional to the square of the distance between them. The force we know as weight is an example of this force, when one of the objects is the earth and the other is the one we hold in our hand or rest on a scale.

In considering weight one naturally asks, How far away is the earth?—for if every rock and pebble of the whole planet attracts the one we hold, the total weight is the result of all these forces of different magnitudes and acting in different directions. By integral calculus, Newton was able to show that the right recipe is simplicity itself: if the earth (or other planet) is a sphere, the force is directed toward the center and its magnitude is as if all the mass of the earth were concentrated at that point. Very good: the direction that is "down" in Australia is not the same as "down" in New York, but each is toward the center of the earth, assuming it is a sphere. Yet the earth is not a sphere. Its diameter measured through the equator is a few miles greater than that measured from pole to pole. The force, therefore, is not exactly toward the center, and this small discrepancy has immense consequences. The moon's orbit is disturbed, contributing to the strange fluctuations in its motion observed (perhaps) by Stone-Age Britons, and it is the off-center attraction of the moon, reacting back on the earth, that produces the precession of the equinoxes.

All this, and much more, is worked out in the *Principia*, presented in the terms of axioms, lemmas, and theorems that Greek authors had established as the standard form of mathematical exposition. The entire work speaks with the voice of certainty, just as mathematics does, and Newton, being human like many of the rest of us, was not above choosing his arguments to achieve a desired result and selecting, or even changing, the experimental data to support his theories.[5] It was not necessary that he do this. In one book he had founded half a dozen new sciences, and the concepts he introduced are the backbone of physics and dynamical astronomy, even though modern methods and results have improved on his. Physics has not stood still. We know that Newton's theory is an approximation, as are, presumably, the more exact theories of relativity and quanta that have extended the range of dynamical thought in the twentieth century. Newton's dynamical arguments refer, explicitly and implicitly, to the concept of time. We must now see how the attachment of a proper meaning to that word helps to relate theory to experience.

Time in the *Principia*

In this book I am trying to explain how the idea of time fits into a scientific description of the physical world. We have seen some of the difficulties encountered by the earliest Greek philosophers in their preliminary sketches of a natural philosophy without exact laws or the insight that these give into the processes of nature. With Newton we have for the first time a system of ideas that leads to verifiable quantitative conclusions. How do they overcome the old difficulties? To this there are two somewhat different answers, Newton's and a more modern view. I deal with Newton's first.

The first of Newton's laws of motion asserts that if initially at rest, an object with no external force acting on it will remain at rest; if not at rest, it will continue moving in a straight line at exactly constant speed. In space there are no milestones, and so we may well ask, A straight line at constant speed with respect to what? A uniformly moving train will not appear so to me if I observe it from an automobile that accelerates or goes around a curve. Further, the definition of exactly constant speed assumes not merely a good clock but a perfect one, a thing as ideal as the mathematician's perfect circle. How are the absolute standard of straightness and the idealized clock to be brought into this world of ours? Newton's answer is given in the Scholium following the definitions at the start of the *Principia*: they are all about us.

Absolute Space, in its own nature, without regard to any thing external, remains always similar and immovable. Relative Space is some movable dimension or measure of the absolute spaces, which our senses determine by its position to bodies, and which is vulgarly taken for immovable space. . . .

And so instead of absolute places and motions we use relative ones, and that without any inconvenience in common affairs; but in Philosophical disquisitions, we ought to abstract from our senses, and consider things themselves, distinct from what are only sensible measures of them. For it may be that there is no body really at rest, to which the places and motions of others may be referred.

Absolute, True, and Mathematical Time, of itself, and from its own nature flows equably without regard to any thing external, and by another name is called Duration: Relative, Apparent, and Common Time is some sensible and external (whether accurate or unequable) measure of Duration by the means of motion, which is commonly used instead of True time; such as an Hour, a Day, a Month, a Year. . . .

For the natural days are truly unequal, though they are commonly considered as equal, and used for a measure of time: Astronomers correct this inequality for their more accurate deducing of the celestial motions. It may be that there is no such thing as an equable motion, whereby time may be accurately measured. All motions may be accelerated and retarded, but the True or equable progress of Absolute time is liable to no change. The duration or perseverance of the existence of things remains the same, whether the motions are swift or slow, or none at all. . . .

There is not much that can be done with these explanations. They appeared to be necessary for the logical coherence of Newton's physics but there is no way they can be put to experimental test. As will be shown in Appendix 3, physics today has only a shadowy idea of how to do without them.

For Newton, absolute space and time were really there, and in his own day this contention came under heavy criticism.[6] Leibniz in particular proposed definitions that contained fewer assumptions. Why not say that the geometry of space consists merely of the descriptions of all possible spatial relations between solid bodies? This meets the requirements of at least a practical geometry, while at the same time introducing no extraneous hypotheses. The verifiable elements of geometry have only to do with the relative positions and movements of things. The basic notion of such geometry is that of place, defined relative to other bodies. The totality of places is called space. Space is therefore not a thing with properties; it belongs to the realm of ideas, and this permits a certain latitude in defining it. Leibniz gives the example of a genealogical tree in which place indicates family relationship, but in which the ordering is again an ordering of ideas, not people or objects. According to this criticism, Newton's absolute space is a rack to hang dynamical concepts on, not a real space. And similarly, time is nothing but the totality of temporal orderings and relations. It too belongs to the realm of ideas, not things.

All that can be replied to Leibniz's criticism is that his space and time do very well as vehicles for abstract thought and carpenters' or surveyors' geometry but are empty with respect to physics. The mere contemplation of events or geometrical relations is such a small part of our experiences of time and space that it cannot be considered as getting to the heart of them. A science of dynamics, which deals with a far richer body of experience of force and response, cause and effect, requires a more solid structure of hypothesis with regard to space and time than Leibniz provided. Newton's hypotheses are perhaps too metaphysical for modern taste but they were adequate to the requirements placed on them. The next steps were taken much later, by men who belong to the modern age.

Spacetime

An object with a force on it, say a planet orbiting the sun, not only traces out a certain geometrical figure—an ellipse—in its orbit but also varies its speed in an exactly calculable way as the orbit is traversed. To put it in the crazy but intuitive language that physicists use, the planet, in order to decide how to move, must know both where it is located and what time it is. The purpose of Newton's absolute space and time was to supply the background coordinate system of space and time against which, in this crude metaphor, the planet charts its course. To survey a cow pasture or a country it is sufficient to have a theory of spatial geometry. To understand how objects move, something more is needed; today we call it a geometry of spacetime.

The idea of spacetime is latent in the passages from Newton just quoted, but the modern concept dates from Einstein's relativity theory, first published in 1905. I can explain what is new about relativistic spacetime by means of an analogy. Let us define an absolute geometry of space by reference to the stars: north is the direction of a line running over the surface of the earth to the North Pole and so on. But if you ask a New Yorker to point toward the north she will point in a horizontal direction, uptown, and if at the same moment you ask a Londoner, he will point in a different direction. (To see why this is so, look at a globe.) Thus the directions of "north," "east," and "up," though perfectly well defined by an observer in New York or London, are assigned differently by the two observers. The same is true of times, because of the time difference between New York and London, but there is no confusion at all between a distance interval of ten miles and a time interval of ten minutes. In distinguishing absolute space and absolute time Isaac Newton was describing the conceptual world in which, whether or not we think about it, most of us live today.

What Einstein did was to show that the privileged role of time is an illusion, just like the privileged role of the New Yorker's north. In fact, what time it is depends on how you are moving and where you are, and not just in the sense of the difference in local times between New York and London (which can be annihilated by listening to the BBC). We can get away with thinking that time is absolutely separate because differences in the way in which two observers assign numerical values to a certain time interval become appreciable only if the observers are moving in relation to each other at speeds approaching that of light—and in human experience so far, they aren't. Nevertheless, there are things—even if not people—that move that fast, and Einstein's formulas are verified every day by experiment (see Appendix 2).

It was not Einstein but the mathematician Hermann Minkowski who first pointed out that what Einstein had done was to substitute the old distinction between space and time with a new geometry in which they are combined into a single entity, spacetime, and different observers assign different directions just as New Yorkers and Londoners assign different spatial meanings to the word "north." "The views of space and time which I wish to lay before you," said Minkowski in a famous address to German natural scientists and physicians in 1908, "have sprung from the soil of experimental physics, and therein lies their strength. They are radical. Henceforth space by itself, and time by itself, are doomed to fade away into mere shadows, and only a kind of union of the two will preserve an independent reality."[7]

In replacing space and time by spacetime, however, Einstein and Minkowski were only replacing two absolutes with a third. And the necessity for any such absolutes at all had been questioned many years previously by a critically inclined physicist, Ernst Mach, in Vienna. Mach had pointed out that Newton's absolute space coincides with uncanny accuracy with the space defined by the fixed stars as we measure them through a telescope. (Since the stars are not really fixed, it is more exact to define Mach's space with reference to their average motion.) For example, the elliptical orbit calculated for a planet in Newtonian theory is stationary, nonrotating as a whole, with respect to

Newton's absolute space, but it is also observed to be stationary with respect to the average motions of the fixed stars. Either this is a coincidence, Mach wrote in 1872, or else there is a causal connection, and he cautiously suggested that the causal connection is perhaps the more economical hypothesis: the stars exert some actual physical influence on the planet, and this, cooperating with the influence of the much nearer sun, determines its orbit.

Mach's suggestion has proved extremely difficult to embody in any coherent physical theory (see Appendix 3), but it was one of the ideas that suggested Einstein's further development of relativity, the so-called general theory, which he published in 1916. In this theory the geometry of spacetime is no longer an absolute but rather is determined, in a way that is specified precisely in equations, by the material content of space. If we knew all about the material content of the universe—its quantity, spatial distribution, motion, and history—we would be able (if Einstein's theory is right) to understand the entire structure of spacetime; but we don't, and so further speculations depend on simplifying assumptions to plug up the vast gaps in our knowledge. These assumptions are chosen, of course, in accordance with the little we do know, and the degree of concordance of even these primitive theories with what is observed and measured by astronomers is truly astonishing (see Weinberg 1977), but our understanding of the universe and its past and future is still in a primitive state. Appendix 4 shows a little of the kind of conclusions that can be drawn at present.

Time Viewed Timelessly

Although Newton's account of dynamics assumes an undercurrent of flowing time, there is, curiously enough, a way of understanding and describing exactly the same theory that makes no reference at all to ideas of flow. Perhaps the best way of explaining this view is to compare it with geometry. Nothing requires a complicated example, so we use a very simple one involving a right triangle. To find the area of the triangle in figure 2, common sense suggests what school teaches: supply another triangle, with dotted lines or just in the mind. It follows from the definition of area that the area of the whole rectangle is the product of its length and width, so that the area of the right triangle, any right triangle, is

$$A = \frac{1}{2} ab$$

Focus on that word "is" for a moment. The first time we had it in the spotlight someone had said "Alice is in the living room." There was no intention of implying that she is eternally there, or even that she passed more than a moment there; only that she was there at the moment the statement was made. "Alice will be in the living room" says something different. But if I say that the area of a right triangle *will be* half the product of its height and base, one feels that something silly and extraneous has been said. The "is" in the original statement includes "was" and "will be," if you like, but it is surely more reasonable to say that this particular "is" has nothing to do with time, that it is the "is" of Parmenides. The fact that we use the same word, "is," to denote

2. To find the area of a right triangle of sides a and b.

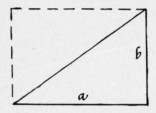

3. To find the distance traveled by an object that starts from rest and accelerates uniformly to a velocity $v = 40$ feet per second in a time $t = 15$ seconds.

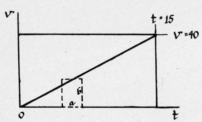

two different relations can be extremely confusing. I will show later that the word "now," closely connected with one of the two meanings of "is" but not with the other, is a very slippery concept requiring careful definition. Somebody once triumphantly defined it for me as "the moment when I am in contact with my surroundings." The question, of course, is what does "am" mean? If it means "am now," then nothing has been defined. But in another sense I *am* in contact with my surroundings as long as my body exists at all. That is a perfectly good use of "am," but it has nothing to do with "now." The same question arises when one thinks about Aristotle's sea battle, in which the difficulty arises from the assumption that every unambiguous statement is either true or false. What does "is" mean? Does it mean "is now"? Clearly that is how Aristotle used it. What happens if we make it the tenseless "is" of mathematics? Can such a proposition have any relevance to human experience? Care must be taken.

To continue, let me pose the kind of problem that you probably had in high school. A car starts from rest and in 15 seconds of uniform acceleration reaches a speed of 40 feet per second. How many feet did it travel in those 15 seconds? Uniform acceleration means that the graph of speed versus time is a straight line. Let us draw it, in figure 3. The thin lines show a construction that will enable us to get the answer: *a* is a very brief time interval somewhere along the longer interval from 0 to 15 seconds, and *b* is the approximate height of the line (approximate because the height changes a little within the rectangle) at that moment. It represents the approximately constant speed of the car during the interval *a*. If the speed during this brief instant is approximately *b*, then the distance traveled during the short interval is approximately *ab*. But this is the *area* of the dotted rectangle. We are now ready to grasp the answer. Fill the whole triangle with little dotted rectangles. The area of each represents the distance traveled during its little interval of time. Therefore the entire area of the triangle equals the entire distance traveled:

$$\text{distance} = \frac{1}{2} \times 40 \times 15 = 300 \text{ feet}$$

or, more generally, if v is the final speed at the end of time t,

$$\text{distance} = \frac{1}{2} vt$$

and the distance is half the product of final speed and time. Now, what kind of "is" have I just written down? Evidently, a geometric "is," the one that was just said to have nothing to do with time. But in this context it clearly has something to do with time: t represents time. I hope it is obvious what to answer: yes, t represents time, but the *relation* between it and the other quantities has nothing to do with time. It is always true.

The Time of Newtonian Dynamics

Perhaps some readers of these crucial paragraphs will have learned something new from them. To those who have plodded toward the obvious with mounting impatience, I say only that the trails of philosophy are strewn with the bleached bones of thinkers who have not grasped that a statement concerned with time need not be made from some particular perspective in time. I shall try to show in chapter 10 what goes wrong when this fact is ignored. For the present, we need only note that statements linking past with present and future need not have any different logical character from statements about the present. To the extent that reality consists of atoms, and that atoms are governed by laws of motion, the mathematical formulation of natural processes is itself timeless, and tells the same kind of timeless truth as any other mathematical description of nature. This fact is of great help in dispelling any feeling that because time relations cannot be seen in the way that spatial relations are—because time "slips away as you try to grasp it"—it has a fundamental nature that is very mysterious. I hope the preceding discussion has shown that in this way of representing them, propositions involving time are not necessarily any more uncertain or hard to understand than those involving space. In neither case do we have the kind of certainty that is usually called "mathematical." If I say that the moon is up and, as evidence, show it to you, I am assuming that the thing we see actually is the moon, that the light has traveled from it to us in a straight line, as it usually does, and very fast, as its habit is. The strength of my plain, ocular evidence rests on many theories that it would ordinarily be a waste of time to question, but theories nevertheless. This is the kind of thing I meant when I said earlier that to be effective practically, logic must be embedded in a matrix of unanalyzed and unstated assumptions. In the present instance they are theories relative to geometry and to the propagation of light. Suppose I now proclaim that the moon will set in two hours. This is

truth of the same kind, depending on the theories just mentioned together with others about the structure and dynamics of the earth-moon
system. None are really worth questioning except in the most frenzied
dispute, and, in that case, all should be. In chapter 10 we shall try to
construct some sentences of the kind that begin "time is. . . ." It will
turn out that definitions of time have a good deal to do with human
purposes, but even in admitting this we must remember that there is
something about the universe that makes it describable by the use of
the dimension of time. What it is about the universe is well exemplified in the laws of dynamics—those of Newton or others more modern.
Just as we need three spatial dimensions to describe our experience of
space, so we need one temporal dimension to describe our experience
of time. Ask a mathematician to construct for you a hypothetical world
of seven spatial and three temporal dimensions and he can do it and tell
you, in mathematical language, what can happen in it. Ask him what
this world would be like in human terms and there is no possible answer. Human terms are invented to describe *our* world; they mesh with
its geometry and its dynamics. Why our world has three dimensions
of space and one of time nobody knows. It may be that the profoundest
laws of nature do not allow anything else, or the question may be like
asking why my car is green (to which one could probably figure out an
answer, but it would contain elements of unimportant fact and in no
way contribute to an understanding of the nature of cars). The question
of dimensions is a truly scientific one, though, and some day there may
be an answer. Whatever this answer may be, the success of the science
of dynamics as a theory that agrees with experience shows us that even
if time has one foot in human senses and imagination, the other is
planted in that nature which we imagine would be what it is even if
there were no humanity on one of its planets; it might perhaps be otherwise but, for whatever reason, it is what it is.

Clocks

5

In old Athens, from the first century B.C., you could step off a street just below the Acropolis into a neat octagonal building and look upon a moving model of the universe.[1] Although the interior fittings are long gone, the building is one of the best preserved of all those that have come down from antiquity, and various sockets and mountings, together with the remains of similar devices elsewhere, have permitted a plausible reconstruction of its inside. Apparently there was a turning dial representing the sky as visible from Athens, arranged so that the observer could see not only that part at present above him but also what was below the horizon. A bronze peg, inserted into one of a circle of holes on the disc and moved every day or so, represented the sun, so that one could see what stars were in the sky when daylight prevented them from being seen as well as the sun's position among them after it had set.

On the roof of the building was a large weathervane. The upper walls were (and still are) decorated with allegorical figures of the winds, which accounts for the building's modern name of Tower of the Winds. Below the figures are wall-mounted sundials.

It is hard for us today to put ourselves in the frame of mind to understand what this building meant to its original users. Remember that most of the names of the constellations that everybody knows are Greek —though of course they surely had other names before these. Remember the highly significant fact that Pegasus does not look like a horse nor Hercules like a strong man nor the Great Bear (our Big Dipper) like a bear, let alone a she bear as the Greek *Arctos Megalé* asserts; everybody is supposed to know that a bear does not have a tail like a squirrel. The constellations and the myths associated with them were part of a very extensive system of ideas that everybody shared. I am not talking necessarily about childlike belief—though that may also have been there. After all, many people today go reverently to churches where they repeat words they do not believe to be literally true. Zeus, the Father of Gods and Men, was a sky-god; his name is related via the Latin *deus* to our word "day." The father of Zeus was the Titan Kronos, whom the old writers often identified with Chronos = Time, and his father, in turn, was Uranos, the sky itself.[2] The sky was not the domain of specialists; there were few specialists and no real conflict between science and religion because the two were, at least in most minds, the same. I hesitate to use the word "astrology," since fortunetelling was the smallest part of the subject, but "astronomy" denotes for us a specialist's domain. Perhaps the word "cosmology" can be taken over to denote a system of ideas in which people *felt* themselves at every moment—but especially outdoors at night—situated among the stars and planets and intimately involved with their motions; and perhaps it is not stretching too far to imagine an Athenian citizen coming out of the sunlight for a moment to orient himself within a cosmos which, for the next few hours, he would be unable to see. Though we today have not much feeling for the unity of nature, for him the world was One.

The Athenian machine was driven by the principle that a water vessel with a hole in the bottom leaks at a constant rate if the water level is kept at a constant height, and that circumstance is easy to arrange. The principle was familiar, since intervals of time were measured by dripping water in many civic functions. In Athenian law courts, for example, lawyers were strictly rationed as to time, and one desiring to speak "asked for the clock."

It is a remarkable thing that amid all this activity nobody seems to have been very much concerned about what time it was. Classical literature refers to time of day only in the most general terms, and one has the impression that in those leisurely times if two gentlemen gave each other rendezvous in the Agora for about the fourth hour, one or the other ordinarily had to wait for a very long time. Civilization itself had to wait two millennia before the truth was finally revealed that time is money. In an interesting paper, de Solla Price points out that the earliest known sundials bore cosmological symbols but no hour lines, that they were used as models of the cosmos, and that timekeeping, when it began at about this epoch, took over a technology that had already been developed for another purpose.[3]

4. Interior mechanism of the Tower of the Winds in Athens, as reconstructed by Noble and Price. The rotating disc carries a diagram of the heavens and is driven by the water clock (clepsydra) shown at the right. By permission of D. J. de Solla Price and Springer-Verlag.

The Tower of the Winds was not, of course, the only device of its kind. There still exist fragments of another by the same maker on the island of Tenos; the Chinese later built huge devices so similar that there may have been some inspiration from the West, and the old clocks that survive from medieval Europe bore moving cosmological models before they had dials. The modern planetarium continues the tradition, and even the dial of a watch can be thought of as a somewhat abstract model in which the position of the hands corresponds to that of the sun. It is only in the last few years that digital watches have begun to displace the heavenly clockwork, and I have noticed that many "friends of time" do not wear them.[4]

What Clocks Measure

What does it mean to measure time, and how is a good measurement to be distinguished from a bad one? Obviously the matter is not simple. To measure the length of a board one lays out a tape, makes sure it is aligned with the two ends of the board, and reads the answer. With time it is different. When you measure a time interval and arrive at its end, the beginning is already vanished into another world—the past—and cannot be revisited. How then can one be sure that the measurement has any meaning? Does it mean anything at all to say that one interval is twice another, or that last year was the same length as this one? Of course it does; the experience of science, navigation, business, and everyday life shows that the world is such that whatever the logical difficulties may at first appear to be, the measurement of time is a reasonable procedure. Perhaps the terminology is not quite right. With the carpenter's tape we are not really measuring length; that is a mental abstraction. Much better to say that we are measuring the board. And similarly, with a clock we are really measuring something about events. The point is that a simple kitchen clock, just as much as the great cosmological models of antiquity, registers the pulse of the universe and keeps time with it, for the same physical laws govern both of them. The quantity t that occurs in these laws is a single quantity, uniquely defined, and it is indifferent, except for questions of accuracy, whether we read it from the clock in the kitchen, the motion of the earth around the sun, or the vibration of atoms in a distant galaxy. Or at least, this is the idea that physicists have. The situation might not be as simple as this, especially with regard to the distant galaxy, but there is as yet no sign of trouble, and the errors in our simple and universal assumption, if any, must be so small as not to make any practical difference in the way we think about time in daily life.

After these preliminaries we can give a first definition of time—not *the* definition, but *a* definition:

Time is what is measured by a clock.

What is a clock?

A clock is a device whose law of motion is known.

Now we have to be a little careful, since one might at first suspect that if clocks involve laws of motion and laws of motion involve time and time involves clocks, a very elementary blunder has been made.

Let us consider a particular clock, a kind used in the Middle Ages: a candle with equally spaced stripes painted around it and burned inside a shield of horn so that drafts would not cause it to burn unevenly. What is the law of motion? One could discourse learnedly about molecular interactions as the candle burns, but the old sexton waiting to ring his bells in the middle of the night knew perfectly well why his candle was a clock: if it is evenly made all the way down and if the stripes are evenly spaced, there is no *reason* why one part should burn more quickly than another. In modern language this is a principle of symmetry; in the seventeenth century it was called the principle of sufficient reason, but anyhow it is obvious and convincing, and it does not presuppose a knowledge of what time is: *all* properties of one segment of the candle are assumed to be the same as those of another. This illustrates an easing of our definition of a clock: we need not know all about the law of motion; it suffices to have the knowledge that makes the device a clock.

There are many kinds of clocks in the world, and most of them can be explained much as the candle clock is explained. How does it happen that we can define time in terms of clocks and clocks in terms of time without running into trouble? There is an assumption of regularity in the world that underlies the definition. We assume, in fact, that there is a universal time that governs all motion. With this assumption it does not matter much whether we actually know the law of motion; we accept that it exists. And the basis for the assumption is our knowledge of the world. It might have been otherwise—at least we can easily imagine it otherwise—but it is not. It is this kind of unspoken understanding of the world that allows scientific ideas to be formulated in rigorous language. Physicists define force in terms of mass and mass in terms of force. The point is that the world is such that this can be done. If it were not, the whole situation would be different and not force and mass at all, nor time and clocks, but other, strange concepts would be needed.

If we were not allowed to argue in this way, time and clocks would have to be explained in terms of other words and these, in turn, in terms of still others, and the process would never end. (All that this proves is that you cannot create a world out of logic, which most people know anyhow.) The category of ideas I am using is what was once called "logic." Now that word is used to denote an abstract line of reasoning that has no necessary connection with the world, a branch of mathematics. Another word is needed for the tool one uses in thinking about nature and I like to call it "universe-assisted logic"; call it what you will, but respect its power.

Good and Bad Clocks

Suppose my drugstore watch disagrees with the time broadcast over the radio. Without much internal conflict, I set my watch. The question arises: What criteria have I for admitting that one clock is better than another? Ultimately, of course, they are criteria based on belief, but they are reasonable. I set my watch because I do not know whether dust has got into it, whether my not winding it till 2 this morning threw it off a little, whether there is some unguessed source of error. On the other hand, though I have not seen the clock in the radio station, I assume that it is a good one, electric, and regulated by a standard time signal. In other words, I concede that the law of motion of the station clock is known better than that of my watch, even though I do not claim personally to know either of them.

Let us take another example. In July of the year 708 B.C. there was an eclipse of the sun, recorded by Ptolemy and by the court annalists of the Chinese Kingdom of Lu. In an eclipse of the sun, the moon casts a small dark shadow on the disc of the earth. *When* the eclipse occurred can be understood from a knowledge of the motions of earth and moon. *Where* it was observed depends on what places on the earth lay in the shadow. Knowing how fast the earth turns and how long ago the eclipse occurred, we can say where those places would have been. Of course, we have to know very accurately how fast the earth turns, since it has turned more than 978,000 times since the eclipse took place, and a small error in the rate of spin is greatly multiplied in the result, but we do know the present rate of spin very accurately, and using it the calculation comes out wrong. There should have been no eclipse in Lu. There are three "clocks" involved here: the spin of the earth, which gives us the length of a day; its orbital motion, which gives us the year; and the motion of the moon, which gives us the month. They have not been keeping the same time.

On different scales of time, we draw the same conclusion from other evidence. Just as a tree shows annual rings, there are corals that under the microscope show daily bands of growth and annual rings as well. How many days are there in a year? Count the number of daily bands in an annual band. It is not easy. The interesting samples are millions of years old and they have been damaged in different ways, but the record is clear enough to show that 345 million years ago, for example, which was in the Devonian era, there were 396 days in a year (Rosenberg and Runcorn 1975).

Modern laboratory measurements are less impressive, perhaps, but more exact. The earth's rotation has been timed for several years by a cesium maser, an atomic clock that is at present the accepted standard of accuracy. The motion has been irregular, but recently the earth has started to lag, and between 1973 and 1975 it lost nearly .02 second. Which clock should we trust: the earth's orbital motion, its spin, the moon, or some atoms of cesium inside a device called a maser?

In order to decide between clocks it is necessary to think carefully about laws of motion and how well we know them. If it were not for the sun and moon, the earth would spin without friction and ought to keep almost perfect time. But they raise tides in the sea and atmosphere, and even in solid land, and the effect of these is to dissipate the earth's spin in friction. So we would *expect* it to slow down, and even though quantitative estimates are difficult, we are not surprised to find that there used to be more days in a year.[5] This implies that we do not expect any change in the earth's orbital motion, which determines the length of a year. Here again, people have looked for a reason for it to change but have found none except for the disturbances caused by the other planets of the solar system. These are very small, and besides, we know the dynamical laws that describe the interaction and can, if necessary, take them into account. The moon's motion is very complicated and consists of several superimposed cyclical changes; in addition, the earth's spin reacts back on the moon to an extent difficult to judge. It is not a very exact clock.

Finally, we have to evaluate the cesium maser as a timekeeper. How it works is unimportant. What matters is that time is kept by the motions of the nuclei of cesium atoms, each protected from the actions of the hostile world by a thick layer of 55 orbiting electrons. The atoms bounce around, but the electrons absorb the shocks and the nuclei swing almost undisturbed, obeying their simple law of motion very precisely.

In summary, we know the laws of motion of some clocks better than others, and when clocks disagree the one we trust takes precedence. It could perhaps happen that two clocks that we claimed to understand very well still disagreed. In that case we would have to conclude either that the concept of a universal time the same for all dynamical laws is wrong or, much more likely, that we aren't as knowledgeable about clocks as we thought we were. But this has not yet happened, at least not seriously enough to make us contemplate any radical alternative to a universal time. Until it does, the unity of clocks, all keeping the same time whatever the physical principles of the mechanisms inside, will be for us a sign of the unity of physics, and we shall continue to believe the great unifying principle: the physical world is such that wherever the quantity known as time occurs in its description, the quantity so denoted is exactly the same.

Clocks and Consciousness

It is premature to confront here the question What is time? since much that will help in the answer remains to be said. But having discussed at some length the simple-minded definition—that which is measured by a clock—we may reasonably stop to ask ourselves what *we* think the clock is measuring. That is, what does the definition say about the human experience of time? No words, of course, are going to define this experience completely, since human experiences are not caught in words, but we can still go much further in relating the time of physical law to that of human life. In Newtonian terms, clocks measure the flow of time. I have pointed out that we do not really sense the flow of time at all, but rather the flow of events, and in these terms the ticks of a clock are merely some more events, controlled, counted, and registered. But whether we speak of the flow of time or that of events, we do sense a flow, and in a certain sense that flow is even. Clocks do not ordinarily appear to us to run sometimes fast and sometimes slowly; their regular ticking seems regular to us as we listen. We do not usually perceive days as being of very different lengths, and when the unusual happens, it is because something has occurred to upset the order of our minds. It is even true that some experimental subjects, awakened at night out of a sound sleep, can after a moment give quite a good guess as to what time it is. There are fascinating books on how animals measure time, and every indication is that the physiological mechanisms underlying these measurements are properly called biological clocks—they share with other familiar clocks the property that all of them appear to be measuring the same thing. Physically, we are made of atoms as was the water-driven clock in the Tower of the Winds. We have seen that it modeled the celestial mechanism in a profounder sense than its designer ever intended, since each may be considered as an embodiment of the same universal dynamical laws. Our experience of time shows that as regards our perception of it, our conscious nature in no way overrides the simple fact that the substance of which we are made is the substance of earth and stars, and moves to the same deep rhythm. This is the meaning of Newton's "absolute, true, and mathematical time" which, "of itself, and from its own nature, flows equably without relation to anything external." He writes this down because events flow equably as perceived by us, and we must understand this as meaning that we are part of the same world that Newton contemplated and are governed by its basic laws. Just as the oldest clocks modeled the known universe, so the Renaissance writers often described man in the same terms. John Donne's fifth Holy Sonnet begins

> I am a little world made cunningly
> Of elements, and an angelike sprite

Man was often called a microcosm (using a word that goes back to Democritus), in an interpretation of nature in which the meaning of every part depended on its correspondence with elements of the celestial hierarchy, the solar system, the body politic, and man. The world, even this recently, was One.

As to the "angelike sprite," we have encountered him before, the ghost in the machine, Descartes' *res cogitans*, the component of the conscious being that resists explanation in material terms. Finally in this study of time we shall have to make our peace with him.

The Paradox
of Reversible Motion

6

Isaac Newton died in 1727, full of years and of honor too, for in the *Principia* he had shown that God, in creating the world, had laid it under the governance of laws of motion from which a man could, on a few sheets of paper, deduce the motions of planets, moons, and comets through the sky. Though the *Principia* contains many other results, this was its main achievement, drawing its glory not only from its climactic relation to what I have called the ancient obsession but from the essential simiplicity of what it revealed. For it is rather simple, once one understands things, to visualize the solar system as a collection of nearly spherical objects, separated by distances so much larger than their diameters that they act effectively as points, and attracting each other with gravitational forces proportional to their masses. Further, the sun's mass is almost half a million times that of the earth (a typical inner planet), so that it exerts gravitational forces on the planets incomparably stronger than those they exert on each other; thus one gets a useful and even simpler picture of the solar system if one ignores the interplanetary forces altogether. The resulting mathematical problem of determining the orbits of planets and comets is then quite easy—undergraduates do it today. The motion of the moon is vastly more complicated, since both the sun and the earth influence it and their relative positions are continually changing; and to "solve the solar system," including the influences of all the planets on each other, is a task that has only recently been attempted with large computers. But it is the technical problems of analysis that are complicated; the layout and the general laws are simple.

Theories of Matter

"God said, 'Let Newton be,' and all was light." Alexander Pope spoke for educated people of the eighteenth century as they felt the shadows fall from their minds. The fragments of old ways of thinking lay around them. Astrology was a perplexing subject. Obviously, the relationship between man and the created universe lay deeper than mere equations, and besides, many people thought that when practiced correctly, it worked. Alchemy also represented far more than the efforts of cranks and mountebanks to make gold. Its practice was with fires and retorts and alembics, but the secrets guarded in its old books concerned the relation of man and nature to the Divine. Astrology and alchemy slipped from sight, not because they had been proved false but because they did not deal with the questions people were asking. What are gases, liquids, and flames? What happens when water freezes, acid bites, wood burns? It was expected—demanded—that the Newtonian clarity illuminate these questions.

Newton believed in atoms, but he had no experimental proof, and conceptually he was scarcely any further along than the Greeks who had preceded him by 2,000 years. There are signs that he intended to elucidate chemistry as he had astronomy; he spent immense effort on chemical and alchemical experiments but finally found little or nothing that added to fundamental understanding. Now we can understand that this was for three good reasons. First, the experimental techniques needed to discover the nature of atoms and their interactions did not exist. Second, he had the wrong mental picture of atomic structure; he envisioned them as very hard little spheres, which stuck together by forces which were unknown but which he brilliantly intuited might have something to do with the forces of electricity or magnetism. (Actually, atoms are complex little structures but the forces are indeed electric.) Third, the dynamical laws he had discovered and applied to planetary systems and laboratory-sized objects break down at the level of atomic smallness and must be replaced by a twentieth-century dynamical theory. This way of thinking, known as quantum mechanics, starts from methodological considerations that come into play when a theory refers to elements of reality on a scale so small that they cannot possibly be the objects of direct experimental study. Its equations, very different from Newton's, can be translated with some loss of meaning into Newtonian language, and when this is done they agree exactly with Newton's equations (or the form they take when the refinements of relativity theory have been added), but at the atomic scale the differences are crucial, and the Newtonian laws do not lead to correct calculations. At this point the situation for the expositor of physics and time becomes a little delicate, since Newtonian language is the everyday language of ordinary realistic common sense, while that of quantum mechanics must be used a lot before one becomes fluent in it and would be out of place in a book of this kind. For the questions to be discussed, the two theories lead to largely the same conclusions and therefore, even though some precision is lost, I am able to use Newtonian language without telling lies. The story I have to tell is subtle enough, even in plain language.

The Picture of a Gas

Imagine a crowd of freely moving atoms or molecules, bouncing off each other and off the walls of their container. This is the physical idea of a gas. We would like to use the laws of dynamics to understand the properties of this gas—its temperature and density and the pressure it exerts. The dominating thought here is the hugeness of a number: one liter of the air in a room contains something like 10^{22} atoms. That is the number consisting of a 1 followed by 22 zeros. If you wonder how anybody can imagine a number so huge, rest assured that nobody can. One just writes down the number. (If people would only learn to visualize a billion, 10^9, I think that national economies would be tighter than they are.)[1]

Although the bounce of atoms at each collision is a relatively simple thing to understand (like billiards in three dimensions), the problem of understanding the behavior of the gas by solving the equations of motion for such a huge number of atoms is inconceivably complicated. But complexity is not the main point, for let us pretend that God has given us a computer for which such a calculation takes only a moment. What would we ask the computer to do? Given the positions and motions of all the molecules at some initial moment, it would be able to tell where they all would be at any subsequent moment. But who can "give" the initial data: who knows it? And who cares? Any such calculation would describe only the behavior of some specific sample of gas, and specificity is misplaced here because scientific questions are general ones: we want to know about the behavior of gases in general. There is a conclusion from this: Newton's laws are not to be used here as they are in astronomy, to project the orbits of each individual moving object, because the situation is different. We see planets but we do not see molecules. If anything is to be calculated by these laws, it must be something of an entirely different character.

What interests us about the solar system is its individual nature. There is, after all, only one of it with its particular set of planets in their particular arrangement in space. There are many samples of gas, and useful scientific statements will concern their behavior on the average, even though no precisely measured sample will have exactly this average behavior. The theory of gases is achieved by marrying dynamics to the science of averages, a union which is called statistical mechanics. The theory is useful exactly because it is general, and it furnishes a prototype of all explanations of the behavior of matter where it is important that huge numbers of atoms and molecules are involved: the arguments must always be statistical, as will be seen especially in the next chapter.

Energy

There is in physics a quantity called "work," which is simply what is expended if one moves an object that resists being moved. Energy is the ability to do work. We buy energy from the electric company and use it to run various electric motors around the house. A motor is something which moves objects that resist being moved. We may also use electric energy to heat the house or light it; nothing is being moved here, but light causes heat where it falls, and heat has the power to do work, as it does in a steam or gasoline engine. Energy is a quantity that is constant through these metamorphoses. That this is so is not obvious; it was discovered slowly, by a number of people during the nineteenth century, that a certain amount of heat is equivalent to a certain amount of work. One calorie[2] of heat is equivalent to 4.19 watt-seconds of energy; that is, the amount of work produced by a one-watt electric motor (if there is such a thing) operating for 4.19 seconds.

In nineteenth-century England, where these facts first became clear, the question of power for mines, factories, and ships was of the first importance. Steam was king, and the science of getting him to work was called thermodynamics. In those days people thinking about energy were conscious of two forms, heat and work, and they formulated the first law of thermodynamics: *When heat gives rise to mechanical work or vice versa, the amount of heat corresponding to a given amount of work is always the same.* That is, heat and work are different aspects of the same fundamental quantity, energy. Later it was realized that there are other forms of energy and it became possible to generalize the law to read, in the second form of the first law: *In any continuous process, the amount of energy entering the process is the same as the amount leaving.* Or, cut to a minimum, the third form of the first law reads: *Energy is conserved.* Clearly there are different forms of energy. Is there any way of looking at the subject so that the equivalence of heat and work, for example, becomes easy to understand?

The key to the equivalence between heat and work is to be found in one particular form of energy, the kind that resides in the head of a hammer that is about to hit a nail and, by driving it a little way, to work on it. This is kinetic energy, from the Greek *kinesis*, "motion." Since a gas consists of nothing but molecules in motion, the heat energy it contains must be in the form of the kinetic energy of the molecules. If a hot gas is allowed to expand and do work in an engine of some kind, it is the impact of the molecules on the moving piston (a force moving something that resists being moved) that delivers energy to the engine. With heat energy thus extracted, the gas becomes cooler; you can put your hand quite near the exhaust pipe of a car and the gas, a second ago at flame heat, will not burn you.

In any simple impact—a molecule bouncing elastically off a containing wall or one molecule colliding with another—kinetic energy is conserved. Thinking again of 10^{22} molecules in a bottle, we see that if all the molecules ever do is collide, and if every individual collision conserves energy, the total energy of the gas remains constant through all the complexity of motion. How do we know that something so simple is true? It is a theorem of Newtonian mechanics. Newton never proved it because he was unaware of this picture of a gas and never thought about this consequence of his equations. But the theorem was latent in them, and almost 200 years later could be incorporated into the theory of gases as it was developed.

Motion Reversed in Time

I have talked about a gas in this discussion because it is easily visualized and the energy is essentially all kinetic. The theory of gases is understood better than that of the other forms of matter, but the remarks to follow should be taken as applicable, with suitable changes, to them as well.

Exchanges of energy with a gas go either way. That is, a hot gas can be made to perform mechanical work in an engine, or on the other hand, if you perform mechanical work on a gas, it gets hot. The familiar example of this is the heating of a bicycle pump when you pump up a tire. This is not just the effect of friction in the pump; the moving piston is speeding up air molecules the way a moving racket speeds up a tennis ball. That the exchange of energy is a reversible process follows at once from the equations of dynamics, for they too are reversible. That is, they make no distinction as to the direction in which time is measured. In the gaseous systems we are considering, the only elementary processes that occur are collisions, and each collision can take place in either direction. Let me say as clearly as possible what this means. If two molecules follow certain paths as they collide, then the collision in which the paths are the same but the velocities are exactly reversed (fig. 5) is also a possible collision. I shall put it another way. Suppose a magic film crew succeeds in taking a moving picture of the collision. I ask for the film and they give it to me loose, in a waste basket. "Sorry, we lost the spool." I assert that there is no possible way for me to decide which is the right way to thread the film into the projector, since either way I project it, the story makes qualitative and quantitative sense.

All that is involved in the changes of material objects is molecular motions. Every elementary interaction between particles of matter is reversible in the sense just defined, and nothing happens that cannot be explained in terms of these reversible interactions; yet the events that take place around us in the world are ordinarily not reversible at all. Give me a film of a child eating a sandwich and I have no difficulty in deciding which end should be fed first into the projector. This is an enormous paradox. How can the result of many elementary processes, each perfect-

5. A simple collision that can take place in one direction can equally well take place along the same paths with the motions exactly reversed.

ly reversible, be a world in which essentially nothing is reversible? The paradox has occupied the attention of brilliant mathematicians and physicists for a century, and great conceptual advances have been made in attempting to resolve it. We know much of the answer, perhaps most of it, but the situation is not yet completely clear and opinions differ on very fundamental points. I can explain here some of the important ideas, but I cannot give the answer in terms that will please everybody.

Obviously it will not do to start off with the child eating the sandwich. The rule in science when confronted with a dilemma is to find the simplest situation in which it occurs and analyze that. Eating is quite a complicated business, and we must start closer to the beginning.

Entropy

Even on a cold day, the molecules of the air outside my house are moving at hundreds of meters per second. There is a huge amount of heat energy in those moving molecules even when the temperature is low. Why should I not be able to persuade some of this energy to move into the house where it would make me warm? The answer is that I can, and domestic heating in the future may be done that way, but it will not happen by itself; I have to make the heat come in. A device that does this is called a heat pump, and almost every house contains one, an electric refrigerator. Put your hand to the air exhaust: it is warm. Part of this heat is produced by the motor; the rest comes from inside the refrigerator. If you want it cold in there you must remove heat, and the refrigerator will not run unless you plug it in. You can't get free heating for your house, or free refrigeration for your lettuce. If you could, there is more you could do: you could run an engine with the heat produced that way and get free power. You can't. This sad fact is embodied in the first statement of the second law of thermodynamics: *It is impossible by means of any self-contained device to cause heat to flow from a cooler body to a warmer one.*

Like the first form of the first law, this one smells of machine oil. We cleansed the first law by introducing a new and general word: "energy." The second law can be similarly generalized by a new word, "entropy," which denotes a concept having just as many ramifications and special cases as energy. Entropy refers to unavailability of energy. No process causes energy to be destroyed, but every spontaneous natural process causes it to become less available. If an active heat source is removed, temperature differences subside; they never increase. If a hammer misses a nail, its energy is still there but it is wasted; it leaks away in heat and sound and no natural process will gather it together to lift the hammer again. If I lift it, I draw on new sources of energy, the food I eat and the air I breathe. Take away these sources of available energy and the hammer rises no more.

The general form of the second law states that every physical process takes place in the direction that makes energy less available. In its second form it states: *Every physical process increases the entropy of the entire participating system.* You say "Wait a minute." The warmth from the sun is continually becoming available for our use. Oil represents available energy, so does Uranium 235, both of which arrived where they are by natural processes. All life (said Henri Bergson) ascends the slope that matter descends. Are there not exceptions to the second law? The answer is that each of them produces a local decrease of entropy somewhere, but none of them takes place in isolation. The fraction of the sun's light that warms the planets is a few parts in a billion; the rest is wasted. The processes that made oil and uranium wasted much more energy than they stored up. Animals breathe out carbon dioxide and neatly dispose of their excreta; our cities produce mountains of trash and in the Los Angeles basin, human activity liberates 20 percent as much heat into the atmosphere as does the sun, all wasted. In every case, the general decrease of availability exceeds the gain. Entropy always increases.

Maxwell's Demon

About a century ago, the Scottish physicist James Clerk Maxwell pointed out that one could imagine a way to circumvent the second law. In a box of gas, install a partition with a little glass door, figure 6. To operate the door there is a demon, gifted with phenomenal powers of sight and muscular coordination, who watches all molecules as they approach the door from either side. He opens the door to allow an unusually fast molecule to pass through it from right to left; then an unusually slow one, from left to right. Otherwise he keeps it closed. By thus sorting out fast molecules from slow ones, he raises the temperature of the gas on the left side and lowers that on the right, so that heat effectively flows from the colder region to the warm one, which is supposed to be forbidden. There is nothing he can do to violate the first law, but it seems that he can violate the second.

In a brilliant analysis half a century later, Leo Szilard pointed out that aside from patience and dexterity, the real commodity the demon needs is information.[3] He needs to know when to open the door. It became clear to Szilard, as it had to a few others, that there is a remarkable linkage between the concept of entropy and that of information. The availability of energy can be expressed in terms of the molecular information that is needed in order to make it available, and the general increase of entropy as the universe evolves and energy becomes less available is analogous to the general increase of randomness and loss of order as anything with moving parts gets shaken around. It is like the shuffling of cards. The letters on this page convey information only because there is a certain order in them. Information is a kind of order.

Having focussed his attention on the information available to the demon, Szilard was easily able to put to rest the demon's apparent ability to violate the second law. If he is a supernatural being, with magic sources of information, he can do it. But if he is merely sharp eyed and fast, he has to find out when to open the door. He must have, for example, a flashlight, and Szilard showed that even the most economical use of a flashlight consistent with getting the required information would increase the world entropy more than the sorting of the gas would decrease it. The idea doesn't work. Details are given in Appendix 1.

6. Maxwell's Demon, an imaginary being who by sorting molecules according to their velocity can apparently violate the second law of thermodynamics.

These considerations have been mentioned in order to show something of the power and universality of this law of nature which gives a one-way character to all physical events. The law, in various domains of application, decrees that rivers shall always flow downhill and never up and that chemical reactions shall progress in the direction they are supposed to go. There can be no doubt that in some way it decrees our own mortality.

If the fundamental dynamic processes that we believe to underly all the general transactions of the world are indifferent to past and future, how can we explain the second law? Generalities are risky. Let us consider a very specific simple case.

One Hundred Molecules

Here is another model. Imagine an elongated box, 1 meter long, containing N molecules of some real gas. N will be, at most, 100. Compared with the several times 10^{23} molecules that would ordinarily be in such a box, this is not many, but we can grasp "one hundred" with the mind, and it will do. The molecules move around the box at speeeds about 100 meters per second, colliding often with the walls and almost never with each other. Collisions with the walls are not perfectly elastic bounces; these are real molecules which sometimes stick for a moment at impact before they jump off the surface again. Their motion is perfectly random. Originally, they were released at the left side of the box. How long before they will all again be in the left side?

If there were only one molecule I would argue as follows: if I look at the molecule at any instant, since it moves at random, I will find it in the left side half the time. But I must play fair. If I find it in the left side and then take 50 more observations in the next 1/10,000 second they will all show it in the left side because it has almost no chance to move in such a short time. If the box is 1 meter long and a typical molecule moves at 100 meters per second, I should wait about 1/100 second between observations in order to give it a chance to redistribute itself. Since on the average every other observation will show it on the left side, I can estimate roughly that it will visit the left side about every 1/50 second.

If there are two molecules there are four possibilities, shown in figure 7. Only the fourth one interests me, and it will occur 1/4 of the time. The general rule is that possibilities multiply. For one molecule there are two possibilities, of which one is looked for. For two molecules there are $2 \times 2 = 4$, for three, $2 \times 2 \times 2 = 8$, for N molecules, 2^N. Therefore before I can expect to see the molecules all on the left side I should look 2^N times, and these observations should be spaced at least $1/100$ second apart. The total time occupied in the search is therefore at least

$$T = \frac{1}{100} \times 2^N \text{ seconds.}$$

It is instructive to evaluate this quantity for several values of N. Suppose there are 10 molecules. The formula then gives about 10 seconds. About this often all the molecules will be (for a moment) at the left (or right) side of the box. If my magic film crew photographed the box for 10 minutes there would be about sixty frames showing all the molecules on the left and an equal number showing them all on the right. The film would be perfectly reversible, and would give no clue as to which end should be fed into the projector. Let us go on:

N	T
10	10 seconds
20	10,500 seconds $= 3$ hours
50	350,000 years
100	4×10^{20} years

What can we make of a number like the last one? The present figure for the age of the universe is about 2×10^{10} years; this happens to be 2×10^{10} times as long. It means nothing to us. The spontaneous gathering of molecules, if it occurred, would make more energy available to

7. The first possible distributions of two molecules in the two sides of a box. All four are assumed equally probable.

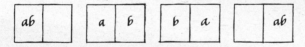

do work. The second law says this never happens. Now our little calculation means something. It means: Exactly what do you mean by never?

There is no sense in even trying to write down the figure that occurs if the box contains a more realistic number of molecules. The point begins to emerge. To see it very clearly, let us look at the experiment from the perspective of Parmenides or the one that simple piety gives to God: "A thousand ages in Thy sight are like an evening gone." On this scale the gas merely fluctuates around; sometimes it is all on one side for a moment, sometimes it is all on the other. A film taken from this august perspective shows the fully reversible behavior that we should expect on learning that the molecules obey reversible laws of motion; if one had time to view the whole film it would reveal no clue as to which end should be threaded into the projector. The second law is therefore false.

Yet of course the law is true. Every power station in the world, every gasoline or diesel engine is designed according to principles that assume the law is true. It is a question of perspective, of scale. The situation is rather novel, and we shall be advised to approach it with some care.

The Causes of Things

What has been shown here for a few gas molecules in a box is generally true. In fact there is a mathematical theorem by Henri Poincaré which states that if the external circumstances are constant and particles are not being created or destroyed, a system of particles that has once been in a certain configuration will, if you wait long enough, finally come back again, as close as you like, to the same configuration. This sounds like a mathematical justification of the cyclic conception of history that underlies so much ancient thought—you can see it in Plato, and many people still think that way (Eliade 1954). But it is not, for the time scales of such repetitions are vastly longer than that of the universe as we understand it, and besides, the return to an original state as described by Poincaré's theorem requires things to happen that are incredibly strange, far stranger than the gathering of particles on one side of a box, whereas the great historical cycles of religious myth are accomplished through a normal march of events. And incidentally, the conditions of Poincaré's theorem are not satisfied in our universe since, especially in its early stages, particles are created and destroyed, and since the whole universe changes as time goes on. The universe does not allow the divine perspective, and we must reckon without the possibility of return.

If the average time between recurrences of the state of the 100 molecules in which all are in the left half of their container is 4×10^{20} years, it does not mean that we must necessarily wait that long for it to happen. The unlikely does happen, after all, and it might happen tomorrow. But suppose someone shows me the results of two observations of the gas. One shows the molecules about evenly distributed and the other shows all of them on the same side. I am asked which was taken first, and immediately choose the second one. My reasoning is that since it is fantastically unlikely that the molecules all got into the same side of the box by themselves, there must almost certainly be some other explanation. My guess is that someone had released them only an instant before, and they have not had time to spread out. I think you will agree I am probably right. Generalizing, we can say that if something is in a low-entropy state it is probably because some human or natural process acted to put it there, not because it got there by itself, and that the natural tendency of entropy is almost always to increase.

"Probably," "natural tendency," "almost always"—these words do not sound as crisp as most of the words of science. There seems to be some weakening of intellectual standards. Further, even if they are taken to be the best that can be done, there remains the paradox with which we started: the second law, even in the watered-down form just given, assigns a direction to events whereas the fundamental laws of dynamics assign none. To repeat, one cannot derive one-directional laws from two-directional principles. Where was the one-directionality smuggled in? I think it was in the assumption that if a thing is in a low-entropy state there is probably a cause at work other than the operation of pure chance. Since everybody knows that causes precede effects, the use of this single word "cause" assumes that time has a direction. But the word settles nothing, of course, since the universe, in operating to produce its causes, can be assumed to be governed by its fully reversible dynamical laws. If a state of low entropy exists it has a cause which can probably be taken back to an antecedent state of low entropy, and so on, back to a first cause, the beginning of it all. As will be shown in chapter 9, there is evidence that the first cause acted at a time not infinitely remote. We *define* the past as the direction along the temporal line in which this event lies. Past-future symmetry is destroyed in this argument just as it is in every event of ordinary life. Note that if there were no evidence that the universe ever had a beginning, if astronomical observation forced us to believe in a universe that either was infinitely old or had an origin so far back in time that no effect traceable to it could be detected, we could not use the words that have just been used. Then what could we say? I don't know; it would depend on what kind of universe it was, how its dynamics operated on a large scale to produce the supplies of energy that allow us to live. Such conjectures are useful because they expose the logical structure of physics, but I do not think they are necessary to an understanding of our own world.[4] In the next chapter we shall see how it is possible to understand the temporal relations of our world in terms of the elementary processes which take place in it.

The Sense
of Past and Future
7

The natural events by which we measure time fall mostly into two categories: those which are fundamentally very simple and those which are very complex. The simple ones are the original clocks of mankind: the earth that spins once a day (as we would say) and orbits the sun once a year. On the level of the finest possible precision these are, as we have seen, subject to subtle and complex perturbations; the errors they introduce are so small that nobody needed to bother about them for thousands of years. The simple motions (ignoring their small perturbations) are governed by the laws of dynamics in particularly simple ways, and in using them as clocks we can claim to know just as much about physical time as we know about the laws. Further away from us in space are innumerable double stars which orbit each other in time to the same laws. In distant galaxies we can no longer follow the motions of double stars but we can follow the same kind of motion on a much smaller scale: galaxies contain hydrogen gas, of which each atom contains one electron and one proton. In these miniature systems, if one makes allowance for differences of description forced upon us by the enormous differences in scale, the same dynamical laws prevail; we know it because we can study the radiations that reach us from these distant atoms and compare them with radiations produced in the laboratory. Wherever these simple motions can be studied in undisturbed, or almost undisturbed purity, we find that a single time variable suffices to account for them all. This world-wide synchronism is expressed in the concept of an absolute time that has the same meaning everywhere, and it is a valid concept even though the theory of relativity (see Appendix 2) shows that the idea must be expressed more carefully than is done here.

All these simple timekeepers are described by equations of motion in which past and future are interchangeable—if we show an astronomer a film of a planet rotating or double star in motion he has no way whatever of knowing whether the film is being shown forwards or backwards.

The other set of physical processes from which we derive our conceptions of time are of an entirely different nature: the growth of children, the changes in our own bodies and minds from hour to hour and from year to year, and, on a larger scale, the gradual changes in the earth's landscape of which we are dimly aware. None of these make accurate clocks; none are simple to understand. All, if we are to trace in them the operation of physical time, require to be analyzed in terms of vast numbers of individual molecular processes, and many of them are still beyond our science. Even though we have every reason to believe that the molecular processes, or better, the motions of the atoms composing these molecules, are finally governed in a simple way by universal laws, the chain of argument is so long, and the number of complicating factors is so vast, that one rarely catches, in the chirping of billions of tiny voices, the music of the spheres. In these processes the direction of time marked *past* is not for a moment confused with that marked *future*. Everywhere, at every instant, a single principle orders the flow of events. It does not control their rate but it decrees their direction in time. It is the second law of thermodynamics.

As we have seen, this law was originally established through studies of heat engines. As scientists began to think in atomic terms (a very different matter from merely believing in the existence of atoms), it began to become clear that the second law describes, in terms of properties accessible to the senses, the consequences of increasing molecular disorganization on a scale far smaller. We do not know the whole story, but apparently it is by tracing the energy transactions of the universe as a whole, in the gravitation-controlled motions that give rise to galaxies and stars and in the atomic and nuclear processes by which gravitational energy finally becomes the energy that sustains the earth, that we shall learn to trace our own genealogy from the single cosmic act of creation.

Scale

If you were to run beside a gas molecule (at some hundreds of meters per second) and ask it whether it is moving at random, it would regard you with surprise. "Do I *look* as though I were moving at random? I am only following the path I was knocked into at my last collision." And of course, viewed at this scale, the motion is not random at all. But observe the motions of 100, or 1000, or 10^{23} such molecules all charging around together and the only reasonable description you can give of the processes is in terms of the laws of chance.

The tossed coin that begins a football game falls not at random, but as it was thrown. If the referee's nervous system were more finely tuned he could control the fall every time. We toss a coin at moments when impartiality is desired because we do not think that from a practical point of view anybody knows in advance how it is going to fall. Randomness is another word for ignorance, and ignorance is often a matter of scale.

Atoms are unimaginable. They are so small that if you were to mark the molecules in a single cup of water and let it evaporate, or throw it into the sea, and wait until winds and currents had mixed it evenly into all the water in the world, a second cup, dipped up anywhere, would probably contain several of the original atoms. A cubic meter of a typical gas under ordinary conditions of temperature and pressure contains about 2.7×10^{25} molecules. This is much more than the number of all the stars in all the galaxies that we can see in our telescopes. We respond to changes of scale by changing words. An individual, or even a family, does not have a birth or death rate; these are terms applied to populations. "Solid," "liquid," "gas," "pressure," "temperature," and "entropy" are words that can be applied only to huge numbers of molecules. Individual molecules move fast or slowly and hit a surface more or less hard; these are the mechanical notions fundamental to temperature and pressure but they do not begin to express the meanings of those two words. The categories of thought originate in our perceptions, and these depend very much on scale. It is important to keep this in mind as we turn to a topic that is central in our understanding of physical time.

Cause and Effect

In order to understand the relations that are described under this heading we must pay careful attention to the scale of the things we are talking about. Let us start at one extreme, that of astronomical bodies. Of course, one can say that the sun's attraction *causes* the earth to move in an elliptical path, but it would be anthropomorphizing the situation to say that in each new second the sun's force is new and the earth's new motion is its immediate effect. In fact the sun and its gravitational field just sit there. Nothing changes, and in no sense has an effect followed a cause. Better to treat the sun and earth as a self-contained system in which the parts are linked by forces and the behavior of the whole is simply described by the laws of motion.[1]

Explaining the appearance of celestial motions by the laws of stellar dynamics, we find no particular need to talk about cause and effect, and I suppose we could look at more and more complicated systems of smaller and smaller particles, until finally we considered the whole universe as a system of cooperating atoms without being forced to change our point of view. But this is not, of course, how we look at things; in fact we don't look at atoms at all. To explain occurrences that arise from the cooperation of a large number of little actions we have simpler and more useful modes of speech. A tree falls into a river *because* the current has undermined the bank. Pointless, because too difficult and conceptually unnecessary, to make an analysis in terms of temporally reversible molecular processes. Instead, we give a perfectly clear description in terms of the temporal sequence of cause and effect. The tree falls because of what the water has done. In daily thought and speech we ignore events at the microscopic scale and describe in large-scale terms the things we notice: the river, the riverbank, the tree. Passing from an atomic description to one in these terms, we change concepts also, and new ones such as cause and effect enter the picture.

If cause and effect are introduced into our thought in this way, what can be said about their temporal succession? We know from experience that effects always follow causes. If a meteor falls from the sky it enters the atmosphere intact. Moving through the air it becomes hot. It leaves behind a trail of heated air and pieces of matter that have broken off. When it reaches the ground, reduced in size, it digs a hole, and slowly cools off by warming the surrounding earth, while the noise generated by its descent is dissipated in the atmosphere. This ramification of effects from a single cause is typical of the flow of natural events, and it is governed by the second law. Clearly the fall of the meteor, like all processes governed by the second law, is irreversible. To reverse it one would have to enlist the help of Maxwell's demon to produce a conspiracy of molecular motions that would focus enough energy on the bit of stone to lift it from the ground and send it shooting back into the sky. In natural processes localized energy and substance tend to diffuse away, dissipated by random molecular motions, just as the shuffling of a deck of cards destroy the traces of the previous game. We know it is very unlikely that shuffling will reestablish the original order. How to talk about randomness in terms that will be useful to explain the facts, and in particular to show why the second law holds, is a delicate question to be taken up again in chapter 9. We shall see there that the passage toward chaos is not quite so simple as it first appears, but as a general tendency the fact is inescapable: mountains wear down; atoms break their chemical bonds and drift apart; we must take care to eat and breathe or else our own bodies will slide down the same slope that the rest of matter descends.

With the example of the meteor in mind to typify irreversible processes, we are in a position to say something about the temporal succession of cause and effect. On the molecular level it is always one event that produces many consequences; never many events conspiring to produce a single consequence. Arguments of this essentially statistical kind persuade us that the processes which our normal concepts and language would lead us to explain with ideas of cause and effect are exactly those in which no premonitions can possibly occur: cause precedes effect, and this is guaranteed finally by the second law. As to why the law is true, we shall inquire in chapter 9.

Portents and Documents

We live surrounded by reminders of the past; of the future we have only our guesses. It is usual to act as though we knew nothing of the future and everything of the past, but this is an exaggeration. Records have gaps, are ambiguous, disappear, and our own clear memories, if we live long enough, often reveal themselves as memories of memories, changing further and further with each remove. I am as nearly certain as I can be that the sun rose yesterday and that it will rise tomorrow, equally sure of both, since both depend on the laws of dynamics, which I trust. Here the situation is simple and I know exactly how the laws apply, but if I compare my knowledge of tomorrow's weather with my knowledge of yesterday's, I enter a realm of almost molecular complexity.

I know yesterday's weather by consulting my memory (or a friend's) or by looking at a newspaper. In my memory the information was coded first on long molecules and then stored in my brain in some other way. In the newspaper, ink is sticking to paper. In either case, the arrangement is stable and will be around for some days; more if special attention is paid to retaining it. Where I live the weather is predicted by people who observe conditions over an area of about 1,000 km to the south, west, and north of where I live, and since air masses travel somewhat less than that in a day, the short forecast is usually pretty accurate. For much longer forecasts, the whole earth must be surveyed. The weather is roughly uniform in a region 100 km on a side and 5 km in height. Appreciable variations begin on a larger scale than this. There are about 15 million such regions in the earth's atmosphere, and a thorough forecast would require this amount of data every day, which is quite unobtainable and would be enormously difficult to process if it were provided. Fifteen million, 1.5×10^7 is not yet of the size of molecular quantities, but it is distinctly larger than the number of pieces of data we normally take into account in trying to plan the coming week. Much less information is needed if I am to tell you about last week's weather. I have only to call a few friends and consult the files of the local paper. The quantitative difference is so great that the situation is qualitatively different, but the difference is at bottom quantitative, nonetheless.

The conclusion of this is that the great difference between our understanding of past and future is a difference in the complexity of the guessing games they offer. At every moment, simple causes are giving rise to complex effects. The past presents relatively few choices—with a small amount of information we can make a good guess as to what it was. For the future, we would need absurd quantities of data to rule out the alternatives. How is this? The amount of data you need before you can say you know something is a measure of the complexity of the alternatives offered. If I say that there is a person in the next room you will not be able to guess who it is. If I say it is a forty-six-year-old veterinarian whom you know well, the possibilities are so reduced that you may be able to guess at once. This is part of the difference between the past and the future. The reason we need so much more data to know what will happen next week is that the number of possibilities then is so much vaster than it was last week—the world will be a far more complex place than it is now, and this increase has been going on since the universe began. At every moment, molecular arrangements that have been locked in place for aeons are being broken up and the molecules are being spread who knows where. The earth is mined and the metal thrown away; its fossil fuels are wastefully extracted and prodigally burnt. Consider a time which may some day come when the energy sources are gone—no more hot stars, no more fuel. The average temperature of the universe will be very low and life will be extinct. At least in limited regions, the entropy will be close to its maximum value. The second law, which commands that entropy never decrease, is not then violated but only irrelevant, since entropy is not increasing much either. The world will remain almost constant in complexity. Then the future will be known almost as well as the past. How can this be? Simply because since nothing will be happening, there will be nothing to remember or predict. It is the tremendous rate at which the world changes now, even our little corner of it, that makes prediction hard. There are other lives and deaths possible for the world than mere equilibrium; a few speculations are in the next two chapters.

The Universe Guessed At

8

It is clear from what has gone before that if we are to support any claim to understanding the irreversibility of events, we must show that we understand something of the history of the universe. I have claimed that the flow of events from past to future was determined once and for all by the event that began the universe, so that every time we take a breath or have a thought, the direction of the processes set in motion is determined by the persisting consequences of that remote event.

Suppose I make some assertion about how the universe began. What kind of scientific evidence is there to support or refute me? The observational data are rather miscellaneous, and consist entirely of radiations reaching us from outer space, which we must detect and interpret. At first, astronomers used only light, and the continued use and development of optical telescopes shows that this source of evidence has not run dry. But there are other forms of radiation that bear the same relation to visible light as high or low musical notes bear to notes in the middle of the keyboard. At the higher frequencies we receive X-rays and gamma rays from the universe. They do not penetrate our atmosphere and so must be studied from balloons or orbiting satellites. These contain special telescopes that can be pointed by signals from the ground, and they send back streams of data to recorders waiting below. At low frequencies, we receive radio waves directly into huge antennas that can pick up signals millions of times fainter than those that enter our radios at home. An important aspect of the information from these signals is that even though they travel at the speed of light, they cover such vast distances that by the time they reach us they carry news not of the present but of the more or less distant past of their origin. As we look outward in space we look backward in time, and the information thus obtained is very useful. To give two examples: first, a decade ago objects were discovered that are quite unlike anything previously known. They are very compact, far smaller than any ordinary galaxy, and yet they emit more radiation than the brightest galaxy. They were named quasars. As more were discovered and their positions plotted in the universe, it was found that they are not evenly distributed in the space we observe, but rather that they seem to stand around the edges of it, so far away that light from most of them takes 10^9 or 10^{10} years to reach us. We deduce that quasars are something that happened at a certain stage of the universe but that they no longer exist. The light in transit has preserved them for us.

The second useful piece of ancient history learned in this way comes from studying in detail the frequencies of the light we receive from quasars and other distant sources. A person who knows music can tell by an instinctive analysis of the tones he hears what instruments are producing them. In the same way, light from a distant star brings us information about the atoms that produced it. The eye cannot analyze as the ear does, but it can be aided by instruments called spectroscopes, and when this is done we can learn much about the atomic mechanisms that produced the light. We learn not only what kinds of atoms they are and the conditions in which they exist but also the dynamical laws obeyed by the electrons inside them. It appears that electrons in atoms obeyed the same laws out there and back then as they do here and now. The structure of natural law is by now so unified and locked together that it is hard to escape the conclusion that the laws of nature were substantially the same 10^{10} years ago as they are today. I do not know how we could have made any progress at all in cosmology, which is the theory of the universe, if we were not able to assume this. But all the same, the universe conceals its mysteries from us and even its extent is unknown. We do not know whether the space we survey is a substantial fraction of it or an infinitesimal nothing. How then can we develop out of this vast ignorance any plausible cosmological theory?

Models

There was once a mountaineer (or perhaps there was not), seeking a cabin to spend the night in, whose trail guide instructed him to walk straight ahead for two hours, to make a right-angle turn to the left, to repeat this process and then repeat it again, and finally to walk for one hour, straight ahead. Mentally, the mountaineer made a map of the directions (fig. 8a). With a laugh, he started off to the left, expecting to reach the cabin in one hour. Instead, he reached the chasm (fig. 8b). It is important to realize that he was not necessarily a fool. His attempt was a gamble which did not pay off, and his reasoning is of a kind almost universally used in science. His mental map is called a model. It represents in simple diagrammatic terms the directions he was given, and it suggests a course of action. When tried, this action reveals that the model has a flaw—it assumes that you can walk along any path you draw in it. The climber's unsuccessful attempt allows the model to be corrected.

The great merit of model building in science is that it allows three processes to occur: it allows experience-based information to be ex-

8a. The mental map, or model, made by a mountaineer reading the description of his route in the trail guide.

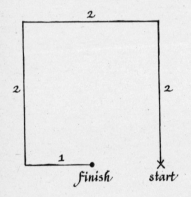

8b. The actual situation represented by the description.

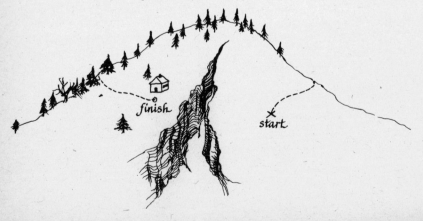

pressed in diagrammatic form so that one can understand the result without having to relive the experience; it suggests new procedures; and finally, a model requires assumptions to be made which, if workable, become parts of our knowledge and if not, have a chance to be corrected. The relevance of these remarks to the study of cosmology is that cosmology as now practiced consists of the making of models. Because we still have not much in the way of data we must assume a lot. New data are coming all the time. At any moment, the existing cosmological models embody the knowledge already at hand. The test of them is to see whether the new data are in harmony with them. For this reason, cosmologists are usually very cautious in making statements about "the universe." Instead, they talk about models, which, since they are creations of the human mind, it is safe to do. The task of cosmology can then be expressed as a simple question: Is the human mind ingenious enough to invent a cosmological model that agrees with observation at all points? If so, that model *is*, in a certain sense, our universe. It will be seen that we are not doing too badly.

Myth and Model

The ancient obsession with astronomy ultimately passed away. There are no stone circles in the British Isles built later than 1600 B.C.[1] By this time, in Mesopotamia, writing had been invented and was being used to record tables of astronomical data as well as ancient cosmological myths. The tables are exact, repetitive, and boring. The myths recount the wanderings and adventures of Enlil, Marduk, Gilgamesh, and other gods representing water, land, sky, and stars (Heidel 1951). There is scarcely a primitive people in the world that does not have its myths, and there are signs that, among other strands, many of these myths contain the tattered fragments of old astronomical knowledge and cosmological conjecture. The great communicative force of myth is used in almost everything Socrates says, and in the *Phaedo*, as he sits in prison with a few friends around him and makes ready to leave the world, his last long discourse contains a myth of the frame and structure of the universe—a fact the more remarkable because Socrates ordinarily showed little interest in scientific theories.

Socrates is the last Greek who makes myth sound real to us. After his time it remains only as fireside tales or as the source of plots and literary references for poetry and drama whose subject was man. The older mythic fragments drift down to us detached from their contexts. There was no story they had to tell that could not be told more clearly in other ways.

Certain commentators on modern science have pointed out that myths are models and models are myths; they have rattled that gourd very loudly, but compare the Hebrew and Babylonian cosmological myths with the cosmologies that originated six centuries before Christ in the Greek settlements along the coast of Asia Minor. Anaximander of Miletus explained the most obvious changes in the cosmos by a model which can be deduced, from the descriptions of later commentators, as that in figure 9. The drum-shaped earth floats in the heavens, surrounded by two circling rings of cloud, which conceal rings of fire. The rings of cloud have holes in their inner surface through which fire can be seen: a round hole for the sun and one of varying shape for the moon. Perhaps I do Anaximander no favor to represent his model by a picture. The sculpture of his period, called Archaic, is more diagrammatic than pictorial, and is not to be taken literally. Greek sculpture became more literal in the fifth century, and so did the cosmological models. The model described by Anaximander requires a large investment of hypothesis be-

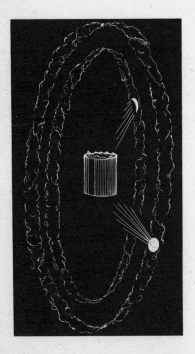

9. The model proposed by Anaxagoras to explain the sun and moon: a drum-shaped earth encircled by two rings of fire concealed in clouds, each fire being visible through a hole in the cloud.

fore it returns a small yield of explanation, but it contains a basic supposition that has stayed with us ever since and has defied all efforts to shake it: the phenomena of the heavens can be explained in terms of mechanisms not entirely foreign to what we see on earth. We have seen smoke rings; we have seen a flame visible for a moment through a rent in heavy smoke, and we are ready to believe that even though we ourselves may not understand how the smoke and flame came into being and move about, they are the kind of thing that can ultimately be explained in plain, ordinary language once we understand them.

How did the universe begin? Many myths answer the question in rather similar terms; their similarity suggests a common experience of tribe or race, or perhaps only the experience of each hunter or farmer trying to hold the outside world at bay for a little while so that he could live his life. These myths appear in many places: in Babylonian and Chinese literature, in Genesis, in Hesiod's *Theogony*, and in Plato's *Timaeus*. Initially all is watery chaos, not in the sense of chaotic movement but in the sense of a perfectly changeless lack of organization. Probably this is fresh water. It never rains salt water, and since it has always rained, the heavens must contain an infinite supply of fresh. These are "the waters which were above the firmament" in Genesis. Then comes a Divine act, which more resembles separation and organization than it does creation out of nothing, and the world begins.[2]

If it did not happen this way, then how did it start? There have been brilliant insights into the history of the universe, but before any reasonable and consistent theory could emerge there had first to be a model of its structure.

Gradually, a store of convincing structural deductions has accumulated. Aristotle argued reasonably that the earth is a sphere because it casts a round shadow in eclipsing the moon: a sphere is the only shape that always casts such a shadow. Further, he said, "it is of no great size," since a few days' journey north or south takes us far enough over the bulge to show us stars we could not see before. (Arguing perhaps from such stellar observations he gives a figure about 60 percent larger than actuality.)[3] In about 280 B.C. Aristarchus of Samos proposed a simple model of the solar system in which the sun stands still and the planets circle about it. Archimedes of Syracuse at once pointed out a difficulty of the new model, which can be explained in this way: if I walk in a circle and look at a nearby tree, the direction from me to the tree changes continually; if I look at the stars from an earth moving in a circle and they do not appear to change positions in the sky, then their distance from me must be unimaginably vast, even as compared with the already unimaginable dimensions of the solar system. This is nothing but the simple truth: the few nearest stars are about 2 million times as far away from us as the sun, and until people could get used to that fact, the Aristarchan idea—even when carefully and thoroughly argued by Copernicus in 1543—raised more doubts in thoughtful people than could be explained merely by medieval thunders from Rome.

Galaxies

In 1750 Thomas Wright of Durham, a man not otherwise widely known, published *An Original Theory or New Hypothesis of the Universe*, in which an explanation was given for the Milky Way. As early as 1610, Galileo had turned his new telescope on it and observed that it seemed to consist of stars, but the idea of celestial spheres was still so strong that nobody before Wright seems to have seriously considered that the stars might be extended in depth. He modeled the universe as a great flat layer of stars with empty space on each side. We are about in the center of the layer; when we look in any direction in its plane (fig. 10) we see innumerable stars, mostly very distant and very close together. When we look at right angles to the plane, they are nearer and fewer. He also had ideas about the form of systems in which such effects of perspective would occur; figure 11 shows one of them.

Five years after Wright, Immanuel Kant seized this idea and in his *Cosmogony* (1900) pointed to the nebulae ("little clouds"), visible through the telescope as pale circular or elliptical patches. He proposed that what are seen are really disc-shaped

> systems of many stars, whose distance presents them in such a narrow space that the light which is individually imperceptible from each of them reaches us, on account of their immense multitude, in a uniform pale glimmer. Their analogy with the stellar system in which we find ourselves, their shape, which is just what it ought to be according to our theory, the feebleness of their light which demands a presupposed infinite distance: all of this is in perfect harmony with the view that these elliptical figures are just universes and, so to speak, Milky Ways. . . .

10. Plate from Thomas Wright 1750 explaining the appearance of the Milky Way as the effect of the earth's being situated at about the center of a vast distribution of stars bounded by parallel planes. In this model the appearance of the Milky Way is simply an effect of perspective.

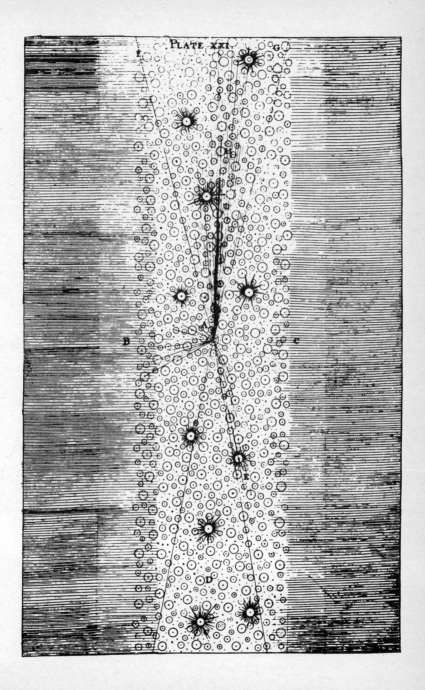

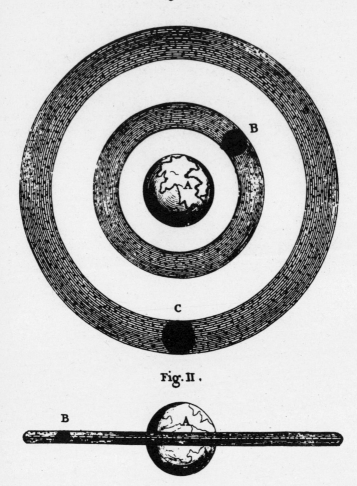

11. A possible form for a "collective body of stars," according to Wright. Because he was concerned that the mutual gravitation of the flat distribution of stars in figure 10 would ultimately cause it to collapse, he set it into orbital motion about a massive central body A "supposed as *incognitum*, without the finite view," so that centrifugal force would oppose gravitational collapse. Figure 10 is now a greatly enlarged view of a segment such as B or C. When we look at this figure we should remember that in 1750 no such sight as those shown in figure 12 had ever been seen in a telescope.

In 1789 Sir William Herschel's great telescope at Slough with its 48-inch mirror permitted him to support Wright's conjecture with actual star counts in the Milky Way, and in 1924 Edwin Hubble, using the 100-inch telescope at Mount Wilson, was able to distinguish individual stars in the nearby Andromeda nebula. With this discovery Kant's hypothesis was established; we begin to have some notion of where we are.

Before Europeans could begin to imagine a history for the universe they had first to free themselves from the limits imposed by the biblical account and its very short time span. This was accomplished during the nineteenth century largely through the work of the Scottish geologists James Hutton and Charles Lyell. They argued persuasively from the known kinds of rocks to the modes of their formation, arriving at time scales in the billion-year range, vastly longer than allowed by orthodox opinion (Toulmin and Goodfield 1965). And finally, before anyone could construct a model of a developing universe to tell us *when* we are, there had to be at least a hint from observational astronomy as to what could be assumed. "In constructing a model," Aristotle once wrote, "we may assume what we please but should avoid the impossible." [4] The task was to guess what is possible, allowed both by what we know of dynamical principles that might have governed its evolution and by telescopic observations telling something of its present state.

A Model of a Changing Universe

9

From the ancient preliterate world we inherit only language, a few artifacts, and fragments of myth. Some stones and some of the myths seem to show that over a long period, celestial motions were observed and remembered. It is impossible to reconstruct the cosmological frame of mind in which this was done, but it seems to have embraced man together with all nature in a cosmic harmony of being.[1] Purified of anecdote, its vastness can be descried in Parmenides, but here man is no longer to be seen. As the underlying logic was forgotten, the gods turned from precisely timed celestial journeys to more mundane pursuits; cosmic ballet turned into opera and then the curtain went down. But still today there may be a fragment left of the old feeling of unity, perceived as the lights fade in a planetarium, that theater where stars and planets act out their parts, or at night, beside a dying campfire. What is the universe, and how are we contained in it?

The Universe Moves

Attempts to convey an idea of the structure of the universe are impeded by the question of its scale. The difficulty is that we tend to measure it in terms of the human arm. It is better if we allow the universe to define its own scale. Galaxies tend to be of roughly the same size and, except for some clusters, spread out roughly evenly in space. They are very numerous. An area of sky the size of the full moon contains thousands of them, so that only a small fraction have been studied in any detail.

Figure 12 shows two typical spiral galaxies. Since they have diffuse edges it is hard to give them a size, but 30,000 light-years might be a typical figure for a diameter that includes most of a galaxy's stars. This is about 2×10^9 (2 billion) times the distance from the earth to the sun, and we are acquainted with the sun, which is 8 light-minutes away. Perhaps we can begin to grasp the scale. These numbers connect the astronomical scale to a scale that is at least related to our experience. But take the galactic size as a unit of distance. The typical spacing between galaxies is about 100 such units, and the distance to the furthest objects studied is about 250,000 of them, or 2,500 times the mean spacing. The first basic fact of cosmology is that, within the limitations of our measurements, the universe seems, on the largest scale, to be roughly uniform. The second fact is that the whole system is expanding. When Vesto Slipher of the Lowell Observatory announced this fact to a meeting of astronomers in 1914 he got a standing ovation. Even though nobody knew where the argument would lead, everyone could see it was one of the greatest discoveries. The result is so fundamental, and was so contrary to the expectations of most people at the time, that I shall give here the three pieces of evidence that now (in the absence of any contrary evidence) make it one of the solidest facts of cosmology.

1. *The Doppler Effect* If a car passes you with its horn blowing, you notice a slight drop in pitch as it goes by. The change in frequency is proportional to the speed of the car and could be used to measure it. The same thing happens with light from a moving source. If the source moves away from us, all colors are shifted toward the red end of the spectrum. In light or sound, this is known as the Doppler effect. If we know what atoms are producing the light and believe that they radiated just the same frequencies in a galaxy millions of light-years away from us as they do on earth, then we can measure the speed of the galaxy. It is easy to know the atoms. Astronomical objects are mostly made of hydrogen, and hydrogen gas emits a light that is impossible to confuse with any other. The few other rays that can be clearly measured in distant objects are also easy to identify. Knowing what these frequencies are as measured in terrestrial sources, and believing, for reasons already explained, that the physical laws describing atoms in different galaxies are no different from those we know on earth, we can determine the Doppler shifts and, from them, the speeds of recession. The light from almost every galaxy we see is reddened, and in distant galaxies the effect is very pronounced. It appears that the speed is proportional to the distance. Of course, we

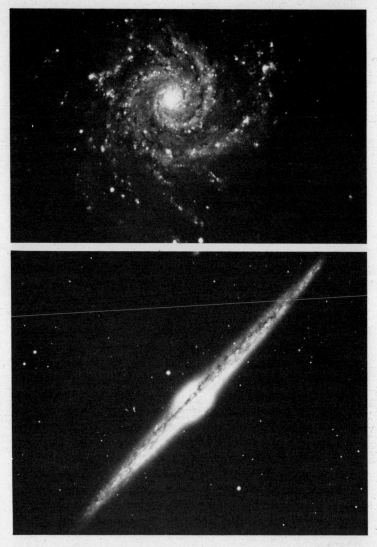

12a. The spiral galaxy NGC 628 in Pisces. It contains roughly 10^{11} stars and rotates clockwise. Palomar Observatory, California Institute of Technology.

12b. Another galaxy, NGC 4565 in Coma Berenices, seen edgewise. This should be compared with fig 11. The obscuring dust visible in the galactic plane extends throughout the galaxy and it is what, in our galaxy, renders the central condensation "incognitum." Palomar Observatory, California Institute of Technology.

do not know from laboratory experience what happens to light if it travels for millions of years en route. It might just redden by itself. To this objection there is no answer except our confidence in a general understanding of the laws of nature. These laws say that light should retain its color. We proceed to the second piece of evidence.

2. *The Background Radiation* A very hot wire glows yellow. As it cools, the color becomes fainter and redder. It is not a pure color but a mixture, and it is the average color that gets redder. (This has nothing to do with the reddening of light from distant galaxies.) Quantitatively, the average frequency of the light changes proportionally to the temperature of the wire measured from absolute zero, while the total light energy emitted changes much more rapidly, as the fourth power of the temperature. Such light is called thermal radiation. An ice cube radiates very little, and at a very low frequency, but it radiates. In 1965 Arno Penzias and Robert Wilson of the Bell Telephone Laboratories, using a receiver originally designed to detect signals from Telstar satellites, found that the earth is bathed in thermal radiation coming from every direction of outer space. The average frequency sounds high—10^{11} vibrations per second—but it is a thousand times less than optical frequencies, and the temperature corresponding to it has been established as about 3 degrees above absolute zero. Many explanations of this radiation have been put forward, but the only one which has held up is that it is radiation left over from an early epoch when the whole universe was enormously compressed and enormously hot. It follows from the laws of radiation that if radiation initially hot is contained in an expanding region it will cool down and become just like the radiation from a cooler object. The background radiation, qualitatively and quantitatively, is consistent with an expanding universe.

3. *The Chemical Composition of the Universe* Wherever there is hot matter in the universe, it radiates. Stars and gas clouds send us their characteristic frequencies, and from these we learn by spectroscopy what chemical elements and how much of each are present in them. It is found that most of the matter in the universe is gaseous hydrogen, the simplest substance, and it has presumably remained unchanged since the beginning of the universe. Most of the rest is helium, and there is a little deuterium gas. Of course, there is other matter—we are, after all, sitting here—but it is not thought to be primordial. Almost all the heavy atoms in us and the objects around us were cooked inside a star that afterwards exploded, spewing them out to drift in space until they gathered together to form the sun and its planets.

When matter is extremely hot, as it must have been in the first stages of expansion, hydrogen is the first element that is formed. Its nucleus consists of a single proton. Deuterium and helium are formed by the collision and sticking together of nuclear particles (a deuterium nucleus consists of two particles and a helium nucleus of four) and the relative amounts of the three gases depend very critically on conditions in the early universe during the short time when they were being formed. At present, about 75 percent of the atoms in the universe are hydrogen, 25 percent helium, .003 percent deuterium, and the rest miscellaneous, mostly carbon and oxygen. Careful consideration of the nuclear processes involved makes it possible to explain these figures provided we assume that the universe expanded rapidly from a state in which it was very hot and compressed.

Thus we have three independent reasons for believing that the universe is expanding: observation of its present state, which seems to *show* it expanding; observation of the background radiation, which seems to show it *has* expanded, and counts of nuclei at present in the universe, which seem to show that it was once very small. The phrase "big bang" to denote the stupendous burst of energy in which the cosmos began has always seemed to me derogatory. In recent years a quantitative theory has emerged which incorporates all the individual processes into a single story with numbers attached. In a bit of mild word-play, this has come to be known as the canonical big bang.

Curved Space

What follows will bear the same relation to the original work as would a thousand-word abridgement of Gibbon's *Decline and Fall*. Of course, the general shape of the events in either work is impressive, but what implants each in our minds is its detail. The recitation of results gives no more idea of science than a summary of events does of history, and it is this sad truth that moves science further and further from the modern consciousness, even as it becomes more and more worthy of being appreciated and understood.

Suppose there is an explosion at time zero, and many particles (atoms? stars? galaxies?) start out from it at varying velocities v. If at the present epoch, time T, we take a snapshot of the whole affair, we see that each particle has gone a distance

$$d = v\,T$$

from its starting point. Therefore if at time T we stand at the starting point and measure the velocities of particles moving away from us, we find the velocities proportional to the distances,

$$v = \frac{1}{T} d$$

simply because the fastest particles have traveled furthest. This proportionality of velocity to distance is exactly what is observed. It is not easy to measure galactic distances. The procedure involves large quantities of stellar theory and is vulnerable to error, but since 1928—when Edwin Hubble and Milton Humason of the Mount Wilson Observatory were able to present data on eighteen nearby nebulae, until today, almost every astronomer has become convinced. Hubble's law, that the outward velocities of the extragalactic nebulae increase proportionally to their distance, seems to hold with reasonable accuracy to the outermost regions of explored space. The constant T, roughly the same for all galaxies, can be measured, and it comes out to be about 18 billion (18×10^9) years.

This is a first crude model, and it must be kept in mind during the refinements to follow, for it contains some truth even if on reflection it is implausible; after all, it almost throws us back into the world before Copernicus. It seems to situate our own galaxy at the exact center, the stationary point from whence all the others rush away. Second, and even more implausible, how can we understand why matter is so uniformly distributed in space? Not only do different "particles" start out with different speeds, but the numbers are distributed so that at any moment the space available for our study is evenly filled. Both these difficulties are solved if we look at the model from the more refined point of view provided by the theory of relativity.

Relativity theory is mathematical, and uses ideas that had been gathering momentum among mathematicians for a century before Einstein borrowed them for a physical theory. I shall express the basic proposition of relativistic cosmology by a rough statement—*The universe expands because all space is expanding*—and then try to explain what I mean.

The situation is easy to imagine in two spatial dimensions. Consider the surface of a toy balloon, and let the positions of the nebulae in space be represented by dots painted on the rubber surface. Now, slowly blow up the balloon. As the surface expands, the dots move further apart, and if you situate yourself in imagination on one of the dots, you will find that the more distant any other dot is, the faster it recedes. (You don't need to buy a balloon. Try it in *one* dimension by marking and stretching a rubber band.)

Somehow the properties of expanding rubber copy those of expanding space. But what is space? Conservative good sense calls it an abstraction: as Leibniz said, merely the totality of the spatial relations between solid objects, a product of the human mind. This conception of space is strong enough to carry at least simple versions of geometry, but we have seen that it breaks under physics. Space must have mechanical properties also, as we have seen in chapter 4. There is more to it than geometry.

The beauty of Einstein's theory is that it begins to unify physics and geometry without the need of untestable hypotheses like Newton's absolute space and time. Dynamics, unlike geometry, involves time, and space and time must be combined into a single abstract entity, spacetime, that serves as a unifying causal background for phenomena. What happens, happens visibly to matter, but the mathematical laws governing matter refer to the mathematical concept, spacetime, and Einstein's theory tells what kind of axioms it must satisfy, though not exactly which ones.

By performing mental gymnastics it might be possible to avoid assigning physical properties to spacetime and instead put them into the theory as additional properties of matter. I don't know that anybody has done this, but I am sure that the mathematical structure so fabricated would lack the formal beauty and simplicity of Einstein's theory of spacetime even if the two finally gave exactly the same results.

If space is said to expand, we must of course specify the scale that is used. If for example some theory were to announce that everything expands—yardsticks, human bodies, universe and all—it would be hard to claim that the statement meant anything whatever. That is not the case here. The objects of the astronomical universe are *observed* to spread out with respect to the structures of metal and glass, flesh and bone that do the observing, and the postulated expansion of space is simply an abstract expression of this empirical fact.

As the dotted balloon expands, neighboring points recede from each other at a rate that increases with their initial separation. It does not automatically have to happen this way. Another substance than rubber might do all its expanding in one small region, leaving the points on the remainder of the balloon at unchanged distances from each other. Contained in the analogy of the balloon there is a physical hypothesis as to the behavior of the underlying substance. It imitates Einstein's hypotheses as to the underlying nature of spacetime, and when this is understood the balloon gives a reasonably exact analogy to modern theory.

That the surface of a sphere is finite but unbounded might be confusing at first to two-dimensional inhabitants of the two-dimensional surface whose entire experience was limited to that surface. We would try to explain the situation to them by describing the sphere in our own three-dimensional terms, but having no experience of a third dimension in their daily lives they would regard the discussion as rather abstract. Finally, the more mathematical among them would understand that the introduction of a third dimension, though unnecessary to describe their daily life, is a mathematical trick leading to a simple explanation of what happens when they explore the world. To the less mathematical who asked "Is this third dimension really real?" they might reply "That depends on what you mean by real."

Now consider our situation when faced with a cosmological model of a universe that is unbounded in the three dimensions of experiential space. We can invent geometries that are finite but unbounded, analogous to the surface of the sphere, but there is no one standing around to explain to us from a superior perspective how they are "really" embedded in a space of more dimensions. We have to understand it all from inside the space we know.[2]

It is hard to imagine a space that simply comes to an end at some point, and I don't think that anyone in the modern era has seriously tried, but the surface of a sphere is bounded without ever coming to an end. An unbounded space can be either finite or infinite, and both possibilities exist in the relativistic theory. In either case, the behavior of matter is connected in Einstein's theory with the character of space, and from this connection arises a simple rule governing every model in which matter is distributed evenly as it seems to be in our universe: *If the universe is spatially finite the expansion must some day cease and become a collapse; if it is infinite, the expansion will slow down but never cease.*

It is plain enough that time adds a sort of fourth dimension to our three-dimensional experience of space. But as I have explained earlier, spacetime is not merely the pasting together of space and time; it is a single four-dimensional entity from which dimensions of space and time can be separated out according to the requirements of any particular observer.[3] Appendix 2 shows a little of the mathematics. The theory is very specific and may some day be proved false in some respect, though it has held up for seventy years. As to why there is spacetime, that appears to be a perfectly good scientific question, but nobody knows how to answer it.

The Canonical Model

Thoreau cited the discovery of a trout in the milk as an example of the power of circumstantial evidence, but circumstantial evidence is only more or less strong; it is never absolutely convincing. Scenarios can be invented that get a trout into that milk without the farmer's having stopped at the brook on his way to town. A model may be supported or discredited by evidence, but this is not the same as absolute proof or disproof of a statement of fact. There is only a moderate amount of evidence to be fitted into the canonical model, but it is impressive how well it fits in, and *where* it fits in.

The model asserts with deadpan literality that the universe began at a certain moment. Not at a certain place, since all space was at that moment concentrated at a point. Then, as now, the universe was everywhere. If we insist on asking where the point was, then we must imagine ourselves into a space of higher dimension and make purely conventional statements unrelated to experience. Let's not do that.

An accurate figure for the date of time zero will be available when we have accurate data on distant objects and an accurate theory of the evolution of the universe. Until then, we have only estimates. A naive interpretation of the expanding universe, as I said earlier, gives the age of the universe as about 18 billion years, but this cannot be exact, since it assumes that everything has been moving at constant speed since time zero, and we can be comparatively sure that the tendency of all things to gravitate together has slowed down the expansion as it progressed. Therefore the expansion used to be somewhat faster than it is now, and time zero is somewhat less than 18 billion years ago. As to how much less, there are various fragments of evidence relating to the speed of stellar evolution and mixtures of nuclear species in various rocks, especially meteorites of ancient origin. A reasonable estimate for time zero is 12 to 15 billion years ago. In a few years it may be placed outside that range, but the number will probably not be very different.

What happened after time zero? There are people who discuss seriously what happened in the first millionth of a second, and they are quite right to do so, since we do not know how good our theories are, and it is only by pushing them into an extreme where they end in absurdity or by showing that they conflict with known fact that we can find out. In the canonical model, the first few seconds are occupied with forming the protons, neutrons, and electrons that populate the world we know. At the end of 10 seconds the temperature is several billions of degrees. For the next 10 to 20 thousand years space is filled with radiation and the fragments of atoms, and most of the helium and deuterium that it will ever contain are being formed. This period ends when the temperature has dropped to about 2,000 degrees and electrons and nuclei begin to combine to form stable, electrically neutral atoms. After this the radiation initially present scarcely interacts with atoms any more, and it is this radiation, unchanged except for the cooling that it undergoes as the universe expands, that we know as the 3-degree background radiation. Thus the background radiation brings us news of that epoch. It is very early. If we represent the history of the universe by a line a mile long, the background radiation reports conditions in the sixth inch of it.[4]

During this time and after, galaxies, stars and planets were forming, but the theory is complex and vulnerable to our ignorance of exact numbers. The usual figure given for the age of the solar system is 4.5 billion years, derived from rather indirect arguments. The age of the oldest known rocks is about 3.6 billion years. This is about one-quarter of the way back to time zero, and by then the universe probably looked much as it does now. On the scale of a mile from time zero to now, the earth has been in existence for the last 2,000 feet and there has been life for about the last 1,500. Some form of man has been in existence for about the last 2 feet, and the Neanderthal comes less than half an inch from the end. The first prehistoric cities are 1/20 inch from the end, and modern industrial society dominates the last 1/1000 inch.

What we really know about physics comes from experiments performed in the laboratory. No result depends on a single experiment; our knowledge has rather the form of a net, and consists of thousands of cross-connections. We may not have interpreted the experiments correctly; our language and presuppositions may not be right, but it is hard to believe that an experiment that gives one result today may give another result tomorrow. Most laboratory experiments last from a few seconds to a few days. They deal with objects that can at least be brought inside a large room. On the cosmic scale they are minute. What gives us any confidence that in extrapolating the knowledge of natural law gained in this way to the cool analysis of an entire universe we are not making ridiculous assumptions? It is possible, of course, that we are, but there is evidence that we are not, and the nature of this evidence illuminates the nature of physical law.

The Quality of Our Knowledge

There is, of course, no necessary logical connection between sensation and concept, between experiment and theory. The two are different logical categories; they cannot be connected. All your experience may be one vast hallucination; you cannot prove that it is not. Your consciousness (or mine) may be the only one in the world, and the landscape, the people, the books may be the inventions of one single mind. But we all have the feeling that the ordinary common-sense explanations are better than these, even though in the absence of a logical argument they may be hard to formulate in a clear and convincing way—most of the great philosophers of the West have had their try at it. We shall see in the next chapter that physics too becomes involved in questions of existence whose answers are not obvious to informed common sense. Relativity contains genuine surprises, and quantum phenomena, not discussed here, contain many more. The language of physics assumes that there is a world "out there," that everybody (allowing for individual variations) experiences it in the same way, and that the knowledge accumulated is common knowledge that can be verified by anyone who makes the effort. This knowledge is codified not in lists of individual facts but in general laws that comprehend a multitude of specifics. It is these laws that, applied to situations far more extreme than the situations that allowed them to be formulated, are at the basis of cosmological theory. The question is, how good are they?

There is something about the formulation of physical law that gives the impression that it is rather unfinished. Take for example Newton's law of gravity, which states that between any two masses separated by a distance r, there is a force of attraction proportional to both masses and inversely proportional to the square of the distance between them. Today we write the expression for the force as

$$F = G \ \frac{m_1 m_2}{r^2}$$

where m_1 and m_2 are the values of the two masses and G is a number that in the scale of centimeters, grams, and seconds is given by

$$G = 6.673 \times 10^{-8}$$

Let it be said at once that this law is easy to apply and appears accurate over a wide range of distances. (Einstein's theory gives small corrections, but they are unimportant here.) It works in the laboratory, where r is a few centimeters and the masses are a kilogram or so. It explains, to minute precision, the motions of planets and satellites in the solar system. It applies to galaxies and probably as well to clusters of galaxies. The separation between galaxies in a cluster is about 10^{24} cm, that is, about 10^{24} times the separation of two masses studied in a laboratory. We may say that the law of gravity is verified over a range of 24 orders of magnitude in distance and, comparing galaxies with laboratory objects, about 40 orders of magnitude in mass.[5] We do not know how large the universe is or what its mass is, but it seems reasonable, if the law of gravity is valid over so large a range, to extend its range still further in our speculations.

Now let us look at the constant, G. It is a purely empirical number that is put into the equation. Nothing we know tells us why it has the value it has. We have the feeling that physical theory ought to tell us things like this, and that our theories are therefore incomplete. There are many such numbers in physics that enter the formulas as letters that stand for arbitrary numerical values to be determined by experiment: the mass m_e and electric charge e of the electron, the masses of other elementary particles, the speed of light c, Planck's constant h which gives the size of quanta. In general, their values depend on the scales used, but there are several relations that are independent of scales. The ratio of proton mass to electron mass, for example,

$$\frac{m_p}{m_e} = 1836.1$$

is, we assume, assigned exactly the same value by scientists on other planets. Three other scale-independent ratios are

$$\frac{\hbar c}{e^2} = 137.036 = R_1$$

$$\frac{e^2}{Gm_e^2} = 4.17 \times 10^{42} = R_2$$

$$\frac{T}{e^2/m_e c^3} = 4 \times 10^{40} = R_3$$

where T is, as before, a number of the order of the age of the universe, and $\hbar$ is short for $h/2\pi$.

One of the most urgent questions of all physics is: Why do numbers like these have the values they have and not some other values? If it is permissible to imagine a universe different from our own, one for example with a different number of stars or a different rate of expansion, would these numbers have the same values there? The fact that the universe is changing in time, that certain aspects of it are not the same from moment to moment, gives some insight into these otherwise impenetrable questions. If there is a unity to the universe in the sense that not only does the universe depend upon the laws but also the laws depend upon the universe, then as the universe changes so should the laws.

Cosmic Coincidences

In 1937 the British physicist Paul Dirac[6] pointed to the quasi-equality of the last two numbers given above and asked whether it is perhaps not a coincidence. If it is not, then the fact goes far deeper than anything we understand, since T refers to the age of the universe while e and m_e are constants of the microscopic world. The first of the two approximately equal quantities, R_2, represents the ratio of the electric force between two electrons to the much weaker gravitational force that acts between the same particles, while R_3 is the ratio between the age of the universe and a time typical of the changes that take place within an atom.

If there is a relation between the two ratios, then we may deduce an important conclusion. T is the age of the universe, which obviously gets one year older every year. Thus we have a relation between numbers one of which is changing, and if this is to remain true, at least one of the other numbers in the relation must be changing also: not all the "fundamental constants" are constant.

Dirac's remark has led, in the years since, to a number of very careful measurements. Since T changes by roughly 1 part in 10^{10} per year, this sets the scale for the other changes sought. Geologists have searched the most ancient rocks for signs that physical processes in them have changed since they were laid down, while astronomers have studied the light from distant sources to see whether atoms behaved any differently in the remote epoch when the light from them started out.

The study is long and difficult and depends on a variety of measurements made by physicists, geologists, and astronomers. For a while there was a suspicion that G might vary, but today it is almost universally accepted that nothing changes as quickly as required by Dirac's argument. Thus the changing state of the universe depends on the laws of nature and the numbers that go into them, but the laws and numbers do not depend on the changing state of the universe. This independence, as I have mentioned before, is fundamental to the current cosmological theories. But do we then have to regard the near-equality of the numbers R_2 and R_3 as a pure coincidence?

We know, by now, quite a lot about how the matter in the universe has evolved from primordial hydrogen. The first nuclear processes in the exploding universe formed helium and a little deuterium, but creatures who ask questions about the cosmic numbers need to be made of heavier stuff; they need to contain relatively large proportions of heavy atoms. As I have said, the only processes we know by which such atoms can be formed are nuclear reactions inside stars which afterwards explode. The material they spew out is gradually pulled together by gravitational forces into new generations of stars and planets. Therefore we, whose atoms were once contained within a star, know that the age of the universe is at least one stellar lifetime. Further, the process of stellar explosion and rebirth cannot have been many times repeated, for the hydrogen that fuels the process gives out. Therefore, the universe is only a few stars old. How old is that? The theory of stellar structure allows us to estimate the lifetime possible for a star. Very roughly, it is given by

$$t_{\text{star}} = R_2 \; \frac{e^2}{m_e c^3}$$

where R_2 is the ratio defined on p. 87. The proposal[8] is that as these words are being read the universe must be roughly this old: $t_{\text{star}} = T$. Solving gives

$$R_2 = \frac{T}{e^2/m_e c^3}$$

and this is the same as R_3. Thus the near equality of the two ratios does not say anything profound about natural law as such, but merely refers to the date at which the question is necessarily asked.

Dirac, commenting on this argument in 1961, wrote "On this assumption planets could exist only for a limited period of time. With my assumption they could exist indefinitely in the future and life need never end. There is no decisive argument for deciding between these assumptions. I prefer the one that allows the possibility of endless life." [9] The arguments are not yet decisive but in the 1970s they have become strong. The "constants" of nature seem not to be changing.

But if they remain constant, life need not necessarily cease. In some unpublished lectures, Freeman Dyson has pointed out that as time goes on and usable energy becomes scarcer, living beings may have to begin to hibernate. While they sleep, energy from the universe's diminishing sources can be stored up so as to sustain life when they are awake, at whatever level of expenditure is deemed suitable. If this strategy is followed there need never be a last generation—though hibernation would be longer and longer, and whether the creature so preserved would resemble us, or even be descended from us, it is a bit too early to tell.

I have said nothing about the ratio called R_1 on p. 87. We now know from astronomical observations[10] that it has changed less than 10 percent since the early stages of the universe. But why is it equal to 137? Nobody knows, but Brandon Carter has argued from what is known at present about stellar structure that unless R_1 and R_2 satisfy the relation

$$R_1 \approx R_2^{1/19}$$

(as they approximately do), the universe would not contain any planets under conditions that seem to us to be the most likely to sustain life.[11] If R_1 were a bit too large all stars would be blue giants and there would probably be no planets at all; if it varied the other way all the stars would be red dwarfs and the planets circling them, unless they happened to be very close, would be cold and dark. But here we are, orbiting a very hospitable star, and what is to be concluded? Is it pure chance that the numbers come out so that life is possible? Is there a reason based on the principles of physics? Or is it possible, as some people have suggested, that we must imagine a whole array of universes in which the constants have different values, and we questioners have grown up in one that happens to be able to support us? When what we can expect to observe is restricted by the conditions necessary for our presence as observers, Carter says that the observation is governed by an *anthropic principle*. We have seen that such a principle explains the approximate equality of the ratios R_2 and R_3. We can explain the relation between R_1 and R_2 by an anthropic principle only at the price of an extravagantly wasteful assumption. Nevertheless, Carter writes, if all other attempts to understand the relation fail, we might have to start taking the assumption seriously, even if we do not like it.

The Origin of Irreversibility

I have connected the relation between cause and effect, and the contrast between past and future, with the gradual disorganization of the universe (or of our part of it), typified by the wearing away of mountains and expressed exactly by the second law of thermodynamics. It is time now to take the last step and inquire why these things are so. The argument to follow is a bit detailed and it has taken about 100 years to develop this far. The amount of time is partly because people were at first confused by the vast scope of the problem and partly because the solution simply had to wait for the canonical model. This problem is not one that can be solved by talking about it; you have to work, starting with a specific theory of the universe. No model approaches the real universe in complexity—it is the purpose of a model to simplify—and our conclusion refers to the model and not to the universe, but it seems likely that the essential argument will carry through into more realistic models in the future.

I have loosely described the irreversible processes of nature in terms of meteorites and falling trees—they are processes in which order passes into disorder. It is time now to be more precise about order, and to make a necessary distinction. By a *macroscopic* order I mean an order that is describable in terms of ordinary words referring to the bulk properties of matter. Even though the typical processes of nature may at first appear to be those of breaking up and wearing away, order becoming disorder, this is not always the case. The mountain now slowly eroding was thrust up originally by normal geological processes, and in the formation of sugar crystals in a saturated solution of sugar and water one can watch day by day the emergence of form out of formlessness. There is in fact one general rule on the macroscopic scale: events take place in the direction that liberates heat. The solution with crystals growing in it is actually a little warmer than the surrounding room. But heat is a concept that bridges the microscopic and macroscopic worlds, for it can be understood only as molecular motion, and it is on the microscopic scale that the direction of physical processes is actually determined. Whether the processes of our experience seem to develop from order toward chaos or in the opposite way depends on considerations at the molecular level, and any explanation of the time direction built into the second law must take place there.

Molecular Order and Chaos

Suppose I have a closed cylinder with a partition across it about 4/5 of the way to the top. The top part I fill with oxygen gas; the bottom with nitrogen. At 9 o'clock I open the partition and the gases start to mix. By 11 the gases have thoroughly interpenetrated and the cylinder is full of ordinary air, a mixture of oxygen and nitrogen. It used to contain two identifiable substances; now it contains one.

At the microscopic level, one might assume that the random molecular motions are at least as chaotic at 11 o'clock as they were at 9. But there is a test of this idea that can be made in the mind, if not in the laboratory. Imagine that it is magically possible to reverse the motion of every molecule at 11 o'clock and send it back, at the same speed, along exactly the path from which it just came. The whole gas will then retrace its evolution and at 1 o'clock sharp there will again be oxygen at the top and nitrogen at the bottom. This experiment is easy to visualize if not to perform, and it forces on us a momentous conclusion: that the gas at 11 o'clock remembers everything that has happened to it since 9. The information must have been stored in the gas somehow; otherwise it could not have retraced its evolution exactly. Of course, it is not stored in each individual particle executing its little random zigzags, but rather in the whole: in *correlations* involving the positions and motions of all the particles. Though the gas may have been chaotic to start with, it develops a vast *microscopic* order, which we could test at any moment if we could arrange that miraculous reversal. Please note that we are here using the terms "order" and "disorder" differently from the way they were used in chapter 7, where the subject was cause and effect and order tended always to change to disorder. There is no inconsistency: here we talk of microscopic order and there it was macroscopic. The example just given shows that they change in opposite directions.

The microscopic order I have described is enormously fragile. Make a mistake in reversing a single molecule and the collisions will start to go wrong. The mistake will propagate through the gas and it will never separate out into oxygen and nitrogen. Never? I showed an example in chapter 6 in which 100 molecules sorted themselves out in the course of time, but the time was ridiculously long and here it will be much longer, something like $10^{10^{24}}$ years. I am not interested in waiting around, and that is why I said "never."

In a real cylinder the microscopic order would be broken up in the course of collisions with the walls of the container, for the molecules in the walls would interact with those of the gas to introduce random perturbations. Very well, let it be a magic cylinder from whose walls the gas molecules bounce like perfect tennis balls. Anyhow, the real universe, to which I propose to apply this argument, has no walls.

There is another peculiarity about the reversed motion: it starts at 11 in a state that is microscopically ordered and macroscopically disordered (a homogeneous mixture of gases) and ends at 1 with less microscopic order but more macroscopic order (the gases are now separated out). A state with a sufficient quantity of microscopic order *can* thus evolve in a very surprising way, a way entirely different from what we are accustomed to seeing in nature. The normal succession of processes described macroscopically is governed by the normal direction of microscopic evolution. The normal direction of microscopic evolution is always from microscopic disorder to microscopic order. It is this tendency, universal in nature wherever we study it, that is expressed in the second law of thermodynamics, in the temporal succession of cause and effect, and in the passage of time. Now, can we understand why the tendency exists?

The examples considered above suggest that a system that starts in a state of molecular chaos, i.e., microscopic disorder, will evolve normally, whereas one that starts in a microscopically ordered state, with suitable correlations, may violate the second law. We need to know whether this is always true. General proofs of theorems of this kind are not available yet, but recent mathematical studies of a number of simplified but plausible models have established that the suggestion we have made holds for them. I am not aware of any proof that it holds for the canonical model of the universe, but I am going to assume that it holds there too.

Before going on I should be a little more explicit. Theorems of this kind are derived from the laws of dynamics, which are indifferent to past and future. They do not state "first microscopic chaos, then microscopic order," for that implies a temporal sequence that is not inherent in the dynamical postulates. Rather, the theorem says that if a system is at a certain moment in a purely chaotic state, then at times *either earlier or later* it will be found in a more ordered state. This is enough, for we wish to apply the theorem at the moment the universe began, and the only times to be considered are later ones; we have nothing to say about what precedes the beginning.

The next question is: does the canonical model start in a state of microscopic disorder? In 1976 appeared a paper by Stephen Hawking of

Cambridge University (who carries the weapons of Einstein if anybody now does) entitled "Breakdown of Predictability in Gravitational Collapse."[12] The phrase "gravitational collapse" refers to what happens when a piece of matter, say a cooling star, cannot exert enough outward pressure to withstand the inward pull of its own gravitational field and begins to shrink. Not only does it collapse inward but the gravitational field develops what is called a singularity, a place where certain variables become infinite and the laws of nature are no longer valid. These objects are called black holes, and the failure of natural law at the center means that they infect the universe around them with an essential indeterminacy, having its origin in quantum phenomena, which renders any exact prediction or retrodiction impossible. In the canonical model the universe begins in such a singularity, and here again the theory is explicit. There are no microscopic correlations at the beginning: this means pure disorder. The argument is thus finished: the universe that begins in microscopic chaos evolves toward microscopic order, and we conclude that the canonical model of the universe evolves normally, obeying the second law of thermodynamics at every step.[13]

Because the idea of normal evolution has been tied to so many other ideas by this point, it will be well to write some of them down separately.

1. The amount of microscopic order in the universe increases continuously from zero.

2. The macroscopic order, which is what we are familiar with in daily experience, changes also. Usually, but not always, this is in the direction from heterogeneous to homogeneous, but it always changes in the direction that we are used to and have come to expect.

3. The behavior of the universe is irreversible in the thermodynamic sense: heat flows from warm objects to cold ones; water in a cup evaporates; machines wear out; rivers flow downhill.

4. We are led by the daily experience of these tendencies to say that time flows always in the same direction or, if we are being more careful, that events do—the general meaning is the same.

5. Past and future, as experienced in daily life, are distinguished unambiguously, and we can define the past, without reference to human perceptions, as the direction along the time axis in which the origin of the universe lies.

6. Causes produce their effects in the future and not in the past.

The entire argument is not rigorous, though it has rigorous pieces in it at key points, because the models it deals with are simpler than the world we know, and we are not yet able to make general theorems that apply to wide classes of models. As an example of the difficulty, the entire science of thermodynamics was worked out for substances whose molecules push each other around or, to put it more formally, interact only when they are close together. But atoms produce gravitational fields that act over great distances, and when there are enough atoms, as in a star or a galaxy, gravitation plays a commanding part in events. We know only the rudiments of the thermodynamics of a gravitating system, and that is not even enough to give a general and satisfying version of the second law. Still, it is comforting to a scientist to realize that all the explicit ideas in the foregoing argument are scientific ideas, that at worst they are a little wrong or a little complicated, and that they are susceptible of being clarified and simplified by the same slow processes of crunching and polishing that, in nonrevolutionary times, gradually bring our science into coherence and order.

The End of Time

As the universe expands, the force of gravity, which tends to bind its parts together, slows down the expansion. But as the parts continue to separate the force gets weaker. Will it finally cause the expansion to stop and become a contraction, or will the galaxies finally slip away from each other's influence and each continue on its journey alone? This is one of the most hotly discussed questions in cosmology, precisely because we are close to having the answer. Cosmological theory (Appendix 4) calls our attention to the quantity

$$\frac{8\pi}{3} GdT^2$$

where G is the (known) constant of gravitation, d is the mean density (mass per unit volume) of the universe, and T is (except for a small correction) its age as deduced from observing the departing galaxies. If this quantity, usually called omega (Ω) is greater than 1, we expect that expansion will cease and be followed by collapse. Einstein's equations tell us that in this case the universe (in the canonical model) is finite in space as well as in time. If Ω is less than 1, expansion continues at a rate which eventually becomes constant and the universe, then as now, is infinite in extent and in material content.

Don't we know the answer without making measurements? The mind recoils from infinity. Beyond every galaxy another galaxy! Yet according to the latest measurements this is the case. It seems that Ω

is less than 1. If it were equal to a million or a millionth, there would be no argument, but for some reason it is fairly close to 1: another numerical coincidence, though Appendix 3 sketches some ideas which might some day make us understand that it is not entirely a coincidence, and Appendix 4 suggests a naive explanation. The main uncertainty is in the value of the density d. It is difficult to count the matter in the universe. Distant galaxies are very faint and get hidden behind closer ones. And how massive is a galaxy? We look at closer ones and count stars. Neither galaxies nor stars vary enormously from one to another, and averages can be taken. But are we counting everything? The centers of galaxies are obscured in gas and dust from which emerge radiations and, occasionally, jets of matter suggesting huge explosions (Calder 1970). What is going on in there? Einstein's theory predicts that a large amount of very dense matter collapses into a black hole. Are there huge black holes in galactic nuclei that contribute to their mass without increasing their luminosity? There are signs that this is so. Are there perhaps smaller black holes elsewhere that we cannot detect? It is known that a galaxy is surrounded by an almost spherical halo of dim red stars. How massive are these haloes? Note that every uncertainty acts in the direction that increases the mean density of matter and hence moves Ω closer to 1.

What does the model predict for us if Ω passes 1? It predicts that one day astronomers in our own or another solar system will look out and see that nearer galaxies are, on the average, approaching. The more distant ones will still appear to recede, but that will be because their light has not yet reached us with the news that they have stopped receding. A period of contraction follows and then, collapse into a universe of heat and pressure similar to that which started out. Then what? John Wheeler of Princeton, who thinks about these things, has claimed that question creates the greatest crisis in physics of all time, for it appears that this is a question to which we cannot know the answer.

There are certain impossibilities in physics that do not rest on our primitive techniques or incomplete knowledge but are part of the fabric of things we *do* know. It is impossible to accelerate an object to a speed greater than that of light; nobody will make a perpetual-motion machine. As we follow the universe backward toward its explosive origin or forward toward its (possible) collapse, we approach a point at which the concentration of matter causes the gravitational field to become infinitely strong and the curvature of spacetime, which in Einstein's theory is connected with it, to become infinite also. This is the singularity. There is no way to write down laws of nature that apply to the world in such a state, and we have seen that it is the existence of lawful behavior that permits us a rational and useful concept of time. Time, in this sense, ceases to exist, and we can reason neither backward, before the beginning, nor forward, after the collapse.

Hawking's result does not mean that we can say nothing at all. It may turn out that there are models in which plausible guesses at least can be made, and a few naive calculations[14] have attempted to show what happens as a collapsing universe moves from cycle to cycle. But it seems that the situation now is as it was in the old legends: however the world is to end, it began in chaos, and none can say what was before that.[15]

It seems appropriate to end this chapter with some words of Karl Popper:

> Every solution of a problem raises new unsolved problems; the more so the deeper the original problem and the bolder its solution. The more we learn about the world, and the deeper our learning, the more conscious, specific, and articulate will be our knowledge of what we do not know, our knowledge of our ignorance. For this, indeed, is the main source of our ignorance—the fact that our knowledge can be only finite, while our ignorance must necessarily be infinite.[16]

Can We Finally
Say Anything Sensible?

10

There is something futile about the question What is time?—and something futile about answers to it. "Time" is a word. We ought to use it wisely, but still it means what we decide it means. What should it mean? The purpose of defining a word is to show where it fits into the intellectual scheme of things. In biology, a word like "gene," in music, a word like "rhythm" changes its meaning in the course of time because it is involved in a changing situation. Wise people do not assign meanings arbitrarily. A good scientific definition ought to be a little compendium of what we understand.

Consider the following arguments.

1. Everything we know about life and consciousness shows that the basis is physical. It is a question of atoms and, of course, of intricate and unknown structures and functions, but they pertain to atoms and to nothing else. Our bodies and brains exist largely isolated from their immediate surroundings but connected to them by physical interactions that include certain sensory pathways. The basic dynamical laws governing atoms and sensory pathways are known, and they are causal in their structure. True, there is a statistical indeterminacy, expressed in Heisenberg's principle and embodied in the equations, which gives elementary particles a certain measure of unpredictable randomness, but life is a matter of large molecular structures, not freely moving elementary particles, and in the analysis of these structures most of the statistical indeterminacy averages out. Therefore the behavior of living matter, like that of inanimate matter, is essentially determinate, and the only difference between the two is complexity. It follows that the actions of a human being are no different in kind from those of a steam shovel: what it

will do in the future is determined by its structure in the present and the signals that have been transmitted to it in the past.

2. What I know is the present, and this knowledge, tested at any moment I choose, tells me that I am free. Past, future, and dynamical laws are intellectual constructions, devised in order to serve certain purposes and justified by the success with which they do so. If these intellectual constructions end by telling me that I am not free, then they are wrong, or at least they are being applied out of their proper domain.

These two arguments arrive at opposite conclusions: I am not free, I am free. By elementary logic, one of them should be false. It should be possible to disprove one or both of them by observation, by experiment. This has not happened yet and I am interested in the possibility, which seems to me very likely, that it will never happen. What do we do then? I take this as more than an academic exercise; the choice bears very closely on the nature of man, of ourselves. There was a parallel situation 500 years ago, when Copernicus moved the earth from the center of the universe and placed the sun there. A tremendous ideological battle ensued, precisely because everybody realized that what was at stake were man's place in the universe and the free action of God. Ultimately the planetary question was settled scientifically,[1] and the ideological positions have undergone steady modification as a result. With regard to freedom, we do not have the scientific decision and perhaps will never have it.

Clearly it is at least as important to our view of man to have a rational consensus on human freedom as it was to agree on the motion of the earth. What is involved in the second argument above is not the random or arbitrary act but the act performed in the living present. As Henri Bergson wrote, "We are free when our acts spring from our whole personality, when they express it, when they have that indefinable resemblance to it which one sometimes finds between the artist and his work. It is no use asserting that we are then yielding to the all-powerful influence of our character. Our character is still ourselves. . . ."[2] What is involved here is not (as sometimes alleged) a conflict between a scientific world view and one based on human values. In the first place, the quest for truth about ourselves is motivated by human values. In the second, a scientific proof that we are not free to change our minds would have the form of an experiment in which people were encouraged to act spontaneously but the experimenter was nevertheless able to predict what they would do. No such experiment has been done. I make this obvious remark only because we are sometimes told that "science" tends toward the conclusion that human behavior is predetermined.[3] It is in the nature of theories to claim that A determines B, and I think this has contributed to the confusion surrounding theories of human motivation. There is no reason to predict that the scientific explanation of human thoughts and actions will be deterministic.

Two Kinds of Time

What is the essential point at issue between the two arguments? Of course it can be formulated in various ways, but I think that the most fundamental way, and the one with the longest historical tradition, is in terms of two different views of time. I shall call them Time 1 and Time 2. Time 1 is the time of physical theory, what is represented in the equations of dynamics by the letter t. It is what is registered by a clock. Time 2 is the time of human consciousness. It is related to Time 1 but the relation is not obvious. It is the time that Eliot had in mind when he wrote "all time is eternally present"; it is a present which is eternally fresh.

It would be idle to claim that every statement about time refers to one of the two I have just indicated. Every thinker has his own point of view, but I claim that the history of speculation on time can be put in order if we note the two contrasting tendencies at work. As to why there are two times, I do not know. The arguments with which this chapter begins are fairly commonplace. I could have started with Plato or with modern physics. The same kind of division is visible in each.

Among the ancients, Heraclitus is surely talking about Time 2 when he says "We step and do not step into the same river; we are and are not." For him reality is process, but it is process viewed always from the standpoint of the present. Parmenides is just as firmly situated in Time 1, the time that binds the universe and all its processes into a single whole which is the subject of the timeless verb "is." I have said that Homer wrote in Time 2. *The Great Gatsby* is written in Time 1.

Plato

Plato seems to have been the first to realize that there are two times and attempt to define the relation between them. I mentioned in chapter 3 that there is no obvious place for time in the theory of Ideas, since the Ideas—the formal and perfect models in terms of which knowledge is possible—are themselves as timeless as a mathematical theorem. Plato deals explicitly with time in the *Timaeus*, which has the form of an extended myth concerning the creation and structure of the world, intended to illuminate the nature of Ideas and their relation with experience.

In the myth related by Timaeus the stars and planets are "living beings, divine and everlasting," and they are governed by a universal soul. The world is created by a divine craftsman (to whom Plato often refers simply as "the god") and we are told that the world was initially timeless, not in the sense that nothing moved but that every motion was of circles turning evenly and nothing ever changed. Then:

> When the father who had begotten it perceived that the universe was alive and in motion, a shrine for the eternal gods, he was glad, and in his delight planned to make it still more like its pattern; and as this pattern is an eternal Living Being, he set out to make the universe resemble it in this way too as far as was possible. The nature of the Living Being was eternal, and it was not possible to bestow this attribute fully on the created universe; but he determined to make a moving image of eternity, and so when he ordered the heavens he made in that which we call time an eternal moving image of the eternity which remains forever at one.[4]

I think the meaning of this passage becomes clear if we assume that what is truly real to Plato is the idea of the universe as Parmenides conceived it, as the One. If the One is considered (as I have argued earlier) to consist of the universe together with its entire history, then it is what it is; it does not turn into something else, and since it never changes, "the One has nothing to do with time and does not occupy any stretch of time."[5] There is nothing illogical here, for the situation is exactly that of the spacetime graph of figure 3 which, while containing a representation of temporal behavior, itself has nothing to do with time. That is why I talked about the graph.

The universe so defined belongs to the realm of Ideas, and Ideas have counterparts, however roughly they may be perceived and understood, in the world of sense. The Ideas of a triangle, of a bed, of justice are each embodied in physical forms and activities, one might call them working models, in which we can glimpse their ideal existence. Now, what is the physical form of the universe? If it is to exist in the physical sense, then time, which was only latent in its Idea, must be actualized in the working model, which becomes a moving image of the timeless.

How is time to be measured? Of course, the simple circular motion of the sky gives us a clock, but there is more to it than this—months and years can be distinguished and recorded only if there occur motions that are not simply periodic: "As a result of this plan and purpose of the god for the birth of time, the sun and moon and the five planets as they are called came into being to define and preserve the numbers of time." [6] Now there was not just a clock but also a calendar.

We should remember that Plato's whole account is in the form of a myth. (He calls it a "likely story.") It is a kind of utterance that we are not used to seeing expressed in prose, though it is still familiar in lyric poetry; neither can possibly be translated into ordinary expository form. About 700 years later, St. Augustine tries to take it literally. In Book 11 of the Confessions he asks, What if the stars and planets should stop while a potter's wheel continues to turn? Would there then be no time by which to measure its motion? And the whole point of Joshua's commanding the sun to stand still was so that the battle could meanwhile continue. He concludes that the heavenly bodies are no more responsible for flowing time than the wheel is, and that though God knows things timelessly, all at once, with no "now," the human sense of time has its origin in the mind, not the stars.

Neither Plato nor Augustine mentions two times, though I think that the concept is implicit in the work of both writers. The idea's earliest appearance that I know of is in a commentary on the *Timaeus* written by Proclus,[7] a fifth-century (A.D.) native of Constantinople who studied in Alexandria and settled in Athens, where he probably belonged to the Academy that Plato founded. There was an interval of about eight centuries between Plato and Proclus, somewhat more than the interval between St. Thomas Aquinas and ourselves, but the Platonic tradition was partly intact. We must suppose that Proclus had not only all the Platonic writings and the commentaries on them (he quotes a number now lost) but also the enormous advantage of a still-living oral tradition.

In his analysis of Plato's text, Proclus first attacks the hard question: In what sense is time an image of eternity, which has no time? In the first place the created world does not belong to eternity, but must have some dimension of its existence that corresponds to it. In the second, this dimension is spread out, for whereas the signature of the Ideas is the One, the created world is diverse. Connecting this diversity there is a sequence, to which we give the name of time, and just as the results of the theory of prime numbers, for example, are not directly visible in the number 1, though they are contained in it, so the manifestations of time are only latent in the ideal world. Nothing happens in Heaven.

A second question which we can answer with Proclus's help is: In what sense do the planets serve to allow time to "come into being?" Here Proclus is very explicit. There are two times:

> Thus ends Plato's philosophical explanation of the time which is the single, whole measure of all things, made and set in motion by the god alone. Next, Plato will discuss time as it is manifest in the heavens, pluralized and fragmented in the various motions of the stars. And when we see that the planets, including the sun and moon, are said to have been created so that this secondary time might become visible, we can see what great dignity is given to the single primary time by the philosopher, or rather by the god. . . .
>
> As we have often remarked, things have a twofold nature: the one invisible, unique, simple, and unworldly, and the other visible, multiple, varied, and distributed throughout the world. There are two kinds of energy: the one primordial, immovable, and intellectual, the other secondary, kinetic, and revolving in relation with the intellect. The one is free from cause and effect; the other contains them. If this is so, then time is also twofold. There is a time for heaven and one for earth. The one remains and at the same time proceeds; the other is borne along in motion . . . the revolutions of the planets create days and nights, months and years.[8]

These two times are what I have called Time 1 and Time 2. If Time 1 seems to be described in rather vague terms, remember that though it was meant to serve as "the single, whole measure of all things," in fact nobody knew how all things are to be measured: even at the time of Proclus, none of the dynamical laws that find their natural expression in terms of Time 1 had yet been formulated. They waited for more than a thousand years.

Plato's Program Fulfilled

I have already mentioned Newton's Scholium on Time, in which he contrasts two kinds of time: "absolute, true, and mathematical time," which "of itself, and from its own nature, flows equably without relation to anything external," and "common time, . . . some sensible and external (whether accurate or inequable) measure of duration by means of motion, which is commonly used instead of true time." True time is the ideal time of dynamical law, whose measure is approximated but not realized by any worldly clock; common time is the practical estimate we settle for. It is the time kept by the world about us, from which we derive our sense of the advance of the present into the future. True time is Time 1; common time is Time 1 modified by the experience of Time 2, but when Newton builds concepts of time into his dynamical theory they all belong to Time 1, which has now at last a place in the scheme of exact knowledge that is more than nominal. In fact Newton never refers to his Scholium again, for his entire book is about Time 1. The realities of our uneven measures and sensations of time are tacitly understood: clocks are simply devices that help to put us in touch with Time 1.

The essential dynamical idea in Newton's *Principia* is a simple differential equation. Permit me to write it down:[9]

$$m\,\frac{d^2x_i}{dt^2} = f_i \qquad (i = 1,2,3)$$

In English, this tells how a force f acts to change the velocity of a moving body of mass m. (Since velocity is the rate of change of position and f changes velocity, the second derivative represents the rate of change of the rate of change of position.) The index i indicates that the same relation holds in each of the three dimensions of space. The time symbolized by t is true time, Time 1.

Different physical systems are governed by different forces. In planetary motion, for example, the force is the inverse-square law of gravity, directed toward the sun. Newton's equation, together with the multiplicity of solutions corresponding to the infinite number of different orbits possible around the same sun, constitute a mathematical structure of exceptional simplicity. The orbits are conic sections: circles, ellipses, parabolas, and hyperbolas. Real planets tend to follow such paths but not exactly, for the force is not quite so simple: planets slightly influence each other's motions. The relation between equation and experienced world is the relation between a Platonic Idea and the experienced world, and the precision with which the tiny equation corresponds with this world has filled many minds with a sense of the majesty and goodness of God. Newton's tomb in Westminster Abbey is conspicuous on the left side of the choir. It bears the inscription

Sibi gratulentur mortales
Tale tantumq extitisse
Humani Generis Decus

[Let all mortals rejoice that such a glory to the human race has existed.]
This was not written because Newton invented the reflecting telescope

or the differential calculus, but because he forged a new intellectual link between man and his Maker.

But Newton's equation is not exactly a Platonic Idea, for though the equation itself and its solutions form a single closed, complete mathematical structure, and the verb "equals" in it is the timeless, tenseless "equals" of mathematics, the equation itself contains t, the time. Plato died 2,000 years too early to grasp this possibility, but it fills the gap left by Plato's silence and Proclus's labored explanation: it shows that Time 1, idealized, conceptual time, is the link between the world of sense and the eternal Idea.

The physics of our own century does nothing to change these relations, but it becomes always more Platonic in its style, since it is based more and more on geometrical symmetry, that is, on abstract form. Where relativistic considerations are involved, this becomes the four-dimensional symmetry of spacetime (see Appendix 2), but even dynamical problems that at first sight have nothing to do with geometry have solutions based on the symmetry properties of abstract spaces.[10]

This is the way in which the timeless principles of modern science are expressed in terms of the abstract and general time I have called Time 1. Our final troublesome task is to try to see how this in turn is linked to Time 2, the time of experience.

The Two Times Connected

The differential equation written earlier is a law of nature. A solution of it, say that corresponding to the case when there is no force at all, $f = 0$, is (in one dimension)

$$x = vt$$

It states in mathematical language that the point representing the object referred to travels with constant speed. The symbols refer to our experience of nature in some way, but how? We look around us in the world and do not see mathematics. We see things and events. Time t occurs in the equation. We do not see time. Galileo once wrote that in perceiving nature mathematically we are perceiving it as God does, but in fact we do not perceive it mathematically, for the equation represents nature only if it is taken together with a verbal commentary that tells what the quantities in it have to do with experience. This is not a trivial step, for it is here that the major controversies and unsolved problems of physics are found. There is not much doubt what to do with mathematics: you solve. The process may be easy or hard but in principle it is straightforward, proceeding deductively in a series of steps toward the answer. Occasionally the process is difficult enough to require a flash of creative genius, but though the creative step may be suggested by considerations outside mathematics, it is purely mathematical genius that is needed. When the work is finished no argument is possible.

The most important interpretive propositions are often called *principles*: in our century the principles of relativity, of equivalence, of indeterminacy focus attention on the deepest problems of physics. Here there are no deductive steps, no proofs. We are standing with both feet in Plato's lower world, the world of opinion, trying to find correspondences with the ideal.

Time 1 is what occurs in the equations. You can see it there. The relation $x = vt$ says: tell me the constant velocity v and give me a value of t: I will tell you what x is. It says nothing about how time gets later and later and the distance x always increases. You can put in any succession of values of t—increasing, decreasing, or random—and the corresponding distance comes out. The idea of the flow of time is totally missing. In his Scholium, Newton tells us that true time, Time 1, flows, but the information is inert and useless, never referred to again. It is a label written in the language of the world of sense but it does not stick; it falls off. Proclus is wiser: "Time moves, not in itself but by virtue of the part it plays in the determination and measurement of motions. . . ." [11] Nothing happens in Heaven.

In all the equations of physics there is nothing that corresponds to the idea of an event. This fact leads to an especially strange situation in quantum physics, which deals with the smallest particles of matter. Here Heisenberg's principle of indeterminacy states that exact prediction is impossible; not merely difficult in practice but impossible in principle. The only analysis of nature possible at that level of smallness is in terms of probabilities. Now probability, though a well-defined concept mathematically, has in its practical uses an element of subjectivity; the probability you assign to a certain event depends on what you know, and quantum mechanics therefore deals necessarily with the knowledge available to an observer. Suppose I toss a die onto a table. When it comes to rest, I look at it. As you and I understand events, the crucial moment was the one when the die struck the table. As quantum physics understands it (if one were to use this cumbersome conceptual apparatus to explain so mundane a process) the crucial moment occurs when I look at the die, for it is then that my state of information changes. The only way to introduce the concept of an event into quantum physics seems to be as a subjective category, outside the realm of physical analysis.

Another way of saying the same thing: in the relation $x = vt$, all values of t enter on the same footing. There is nothing in the equation to suggest that one time, *now*, has a special status. Any value of t is now. Now is when events happen. There is no now in Heaven. Time in Heaven is marked by numbers; that is Time 1. Time 2, in the world of events, is marked by events. Now is the focus on these events. It is, in Bergmann's pregnant phrase, "the temporal mode of the experiencing ego."[12] "Temporal" here means what corresponds, in human experience, to the abstract concept embodied in Time 1.

It is tempting to think of events as extended along a time axis, with the past at one end and the future at the other and the present a dot somewhere in the middle, moving from the past towards the future so that events at one moment in the future are, after a little time has elapsed, in the past. This metaphor is a bubble; it bursts when you touch it. Let me touch it: How fast is the point moving along the line? There have been people who answered this by saying that the time passes at the rate of one second per second, but it is hardly polite to answer a serious question with a tautology. The trouble is that time does not pertain to the time scale on the line from past to future any more than weight pertains to the scale on which each morning a dieter plots his weight. Change is represented *on* the graph. A graph which itself changes is a hybrid representation which as far as I can see represents nothing, except perhaps a confusion of Time 1 with Time 2.

The puzzles that arise out of the human consciousness of time can often be rephrased and solved in terms of Time 1, though we who live in Time 2 may not always be comfortable with the solutions. I could give many examples, but three will do.

1. What is special about the instant "now"? Nothing at all. At every instant of my life I am in contact with my surroundings at that instant simply because light and sound travel so quickly to me that there is essentially no delay. Because at any moment my senses are full of the impressions of that moment, it is far more vivid to me *at that moment* than at other moments earlier or later. Now is any time.

2. Why is your now the same as mine? If the question refers to you as the reader, and to me as the writer, it isn't the same; there are months or years of difference. If we are in the same room talking, the quick pas-

sages of sound and light lock our consciousnesses together. If we are not in any sort of contact, how could anyone tell? The question contains an untestable assumption.

3. If all moments have an equal right to be called now, why do we read events from past to future rather than from future to past? But wait a moment. Are we sure we don't just read them? This is a very important point, so let us consider it carefully. Suppose I maintain that the events of my life are ordered by the time quantity only in the same way that the values of x are ordered by the quantity t in the relation $x = vt$. How could you prove to me that I have to read these events, or that I do read them, in a particular sequence? The birthday party at which I spilled chocolate ice cream on my clothes and the mothers mopped it up is "there," on its proper place on my time axis in spacetime. The "is" I have just written is the atemporal "is" that makes no claim that the birthday party is "there now." (The universe pictured as spread out *now* in spacetime is called by some writers the block universe. As far as I can make out, and for reasons I have tried hard to explain, this picture is logically self-contradictory and makes no sense.)

My body is at that remote point in spacetime, with memory and senses fully alive. Do you wish to say that I have "gone back there"? I wouldn't say it, because the phrase assumes a present moment from which one has gone back, but never mind. I am there. I know nothing of my future life; for reasons related to the second law its traces are not in my memory. What is the difference between what you might call a return to the party and my simply being at it? None whatever that I can see. I am there; that is all. At the moment of writing these words the event is vague in my memory because the segment of time axis in between is many years long. Both events are there, not (to repeat) now, but in the atemporal sense. So are the events of my future. Not that the future is predetermined, that I have no choice in the matter. To say that is to think of spacetime as existing *now*, giving time the dual and self-contradictory roles of an axis of spacetime and also what is read by the clock in the consciousness of people who stand, as it were, outside spacetime and look at it. This is the logical error that in these paragraphs I have tried to combat. Although we commonly think of ourselves as moving forward in time (or time moving against us like a river moving against a rock in its center), I can think of no evidence that this is a fact, or even of any way of stating it coherently. Therefore, except for the trivial remark that at any instant we know the past much better than we know the future, I can think of no reason to say that we read events from past to future.

Now I will ask a question: Is time travel in principle (never mind the technical difficulties) a possibility? It has received some thought in the past and deserves some more.

It is good to think occasionally of the world in terms of Time 1 precisely because we normally think in terms of Time 2. Time 1 makes possible the contemplative study of science and history and our own acts. Time 1 is the time of melody and musical structure; Time 2 is the time of rhythm. The visual arts live mostly in Time 2. Time 1 shows up counterfeit questions like the three just discussed. People not used to thinking like physicists, in my experience, find Time 1 baffling; physicists often refuse to believe that Time 2 has any kind of reality. What is the relation between Time 1 and Time 2? It is a limited part of what from the beginning has been a central question of Western philosophy: What logic relates concept to experience?

I do not expect ever to see a beautiful new answer to that question. Plato's is beautiful, and I have tried to show that there is much truth in it, but nobody wants to be told that the Ideas are really there. As I have tried to show, a description of the world in terms of Time 1 lacks the essential element, *now*, that is needed to make sense of Time 2, while Time 2 does not contain the idea of extension that explains Time 1. These two sets of ideas exhaust our clear and simple conceptions of time, and neither contains what is required to make a bridge to the other. The final explanation, when and if it is developed, will therefore have to use more general and abstract language, not rooted in either familiar kind of time, and it may not be very rewarding for the simple seeker after truth. (I believe it will finally depend on the structure and function of the brain.) Of course it is better for us to know than not to know; of course we cannot have a just view of ourselves as human beings if we do not understand our place in nature. Yet when the light of truth breaks around us we begin to appreciate that shadows have their virtues too, that there may be great truths that cannot be clearly expressed. It is good to love fact, but less good to believe uncritically in its ultimate power. The relation between self and what is not self, exemplified in the relation of now to the time of physical science, has been made vivid by myths, but perhaps it will never be clarified by facts. And besides, as I wrote at the beginning of this book, what is the point of my trying to tell you what time is? You already know.

What, Then, Is Time?

At the end, we can hardly help returning to Augustine's question. It still has the same bleak answer: a sound has any meaning we choose to give it. Yet now we are perhaps able to apply some sensible standards to the use of this sound. Words should not be given private meanings, for they exist in order to be shared. So it is not enough to say with Thoreau "Time is the river I go a-fishing in," or with some physicists I know, "It is what is represented by the letter t in the equations of dynamics." What, in perceptual terms, this symbol represents is a real question; if we know the answer we are on the way to a simple definition. Certainly, time is not a thing. I do not mean merely the kind of thing that can be felt and weighed, though Newton's absolute time which "flows equably" almost seems like that kind of thing. In a sense an event is also a thing, a thing that happens. Is time a thing in this sense? I think not. The present happens; surely this is not an abuse of language, but the past and future do not happen; they are in some hard-to-define sense "there."

I have tried to make it seem sufficient to say that events are the markers of time for us, and that the flow of events, inside and around us, is what corresponds to the ordinary metaphor of the "flow of time." As long as events flow, time does not need to flow. What does it need to do? It seems to me that it needs to denote a certain dimension of our experience. No need to flounder trying to explain what dimension it is: that is part of the utility of physics in this discussion. It is the dimension of experience that corresponds to the quantity denoted as time in the laws of dynamics. Now we must go carefully. Time in dynamics corresponds to something in our experience. Something in our experience corresponds to the letter t in dynamics. Is this a circularity? No, because dynamics is anchored in something else. *The physical world is such that it can be described by the equations of dynamics.* We may take this to be a fact, for the equations have been tested, by confronting them with the world, over and over again. This fact must be the anchor of our speculations.

Putting It Together

Our understanding of the particulars of the world about us comes in two stages: there is a nonverbal stage, when we form an idea, and then a verbal stage, when we try to find means to express it. What happens during the two stages is very different.

A kitten opens its eyes for the first time and sees a pattern of light and shape. I think it is safe to assume that at first this pattern means nothing. During the next weeks the kitten learns to use its eyes to avoid bumping into or falling off things and to find what it wants. The process seems to be one of trial, error, and memory. A grown animal has an idea of spatial relations. It looks behind a mirror to find where the image originates and finding nothing there it does not look again. Immanuel Kant would have us believe that for humans this knowledge of space is innate; that no amount of explanation or experience can supply it to the mind that lacks it. There may be some truth to this, but it is clear that experience plays a major part in a kitten's learning about space, for there is a stage during which this learning must take place if it is to happen at all. Experimenters have shown that when the experience is delayed too long, the kitten apparently *sees*, in the sense that images form, but it never *learns*. Thus if there is an innate component to the knowledge of space there is an experiential one also. I think this is true of time as well as of space; of babies as well as kittens.[13] Nevertheless, Kant's suggestion is very powerful: that space and time should be viewed not as particulars, with Newton, nor as relations between particulars, with Leibniz, but as interpretive forms of perception that, however they get into the mind, must exist before any spatial or temporal interpretations of sense data can be made.

The logical process by which sense impressions, interpreted through the forms of perception, first become specific knowledge and then, perhaps, contribute to general knowledge is the subject of Kant's *Critique of Pure Reason*, an immense work whose vocabulary exasperates, whose difficulty seduces and repels; which, like a dangerous shore with a lighthouse on it, affects the course of every voyager who even comes near.[14] We need not come very near, but the beacon shows that we are in sight of land, for at this point a hierarchy of considerations can be distinguished in the attempt to untangle the concept of time. First, at the mute level of babies and animals, time is a form of the mind which helps organize the inchoate signals of sense into the perception of objects and situations. The kind of consciousness to which this leads is what I have called Time 2, the sense that objects move and situations develop: the flow of events, viewed always from the perspective "I, here, now." It is Bergson's *durée* which, he tells us, "is the form which the succession of our conscious states assumes when the ego lets itself *live*, when it refrains from separating its present state from its former states."[15] In this mode of consciousness we are not conscious of time at all; in fact, as soon as we focus attention on the experience of time, we begin to see events against

a linear scale, marked off with hours or dates, on which the consciousness of a particular moment is registered as a point. Another moment is
another point, but do not try to represent the passage of time by endow-
ing this point with motion! The time scale is Time 1, and it is *impossible*
to represent on it the experience we know as the passage of time.[16] But
it represents the time that belongs to historical reflection as well as to the
equations of mathematical physics which (in some way we do not understand) correspond to a real world existing independently of ourselves.
The fact that the physical world is such as to permit us to describe it with
these mathematical structures and not some others legitimates the entire
system of ideas, down to Kant's perceptual form. If spacetime had two
timelike directions, for example, or if the origin of the universe were
more remote so that it did not always impose its order on events, or if
our bodies were much smaller so that the random behavior of nature on
the small scale with its departures from causality were more apparent,
we would have some perceptual form but it would not be the same as the
one we have.

Time 1, which unlocks for us the mysteries of physical process, cannot be used even to represent (let alone explain) the passage of time.
Therefore, remembering the Parmenidean ancestry of Time 1, we can
conclude that if we are to communicate the nature of Time 2 we cannot
consider the universe (which contains us) in its eternal aspect. This is
what Plato meant, but now perhaps we understand him better.

Have we any other experiences of pairs of ideas, each of which con-
tributes to our understanding of the world but both of which cannot be
held at the same time? Free will and determinism are such a pair: the
mere fact that we are conscious of our inner freedom must not blind us to
the fact that the atoms of nature behave like parts of a vast machine. As I
have said, it seems unlikely that a scientific proof will be given that one
or the other of these hypotheses is wrong; we must learn to use each in its
proper sphere of thought. Yet this example, though it may illuminate
our discussion, does not advance it, for free will is simply the aspect of
causation belonging to an understanding of the world in Time 2, while
determinism belongs to Time 1. To find other examples it seems we
must turn to physics, and here not only are they numerous, but there is a
theory that tells how the apparent contradictions are to be avoided.

The theory is known as complementarity. It expresses the fact that
language is not, and must not be, very closely tied to experience, since
it must allow us to generalize and classify at the same time that it de-

scribes and denotes. Let us take an example of complementarity that is by now thoroughly studied and well understood. Since the early nineteenth century it has been known that light is a wave, and dozens of different experiments can be used to show this to be so. It has also been known since the beginning of this century that light consists of particles, and again there are dozens of experiments to prove it. (We know, furthermore, that this apparent duality extends to sound and many other forms of energy.) How can anything be a wave and a particle at the same time? That is not the right question. The wave experiments are designed to ask nature: Is light a wave? and the answer is yes. Similarly with the particle experiments. They are different experiments, and do not require light to be both at the same time. But how can it be both, even at different times? To answer this question the Danish physicist Niels Bohr was led to a profound study of the nature of human knowledge.[17]

To understand the apparent duality we must consider the experimental facts out of which it arose. We all have had experience with waves, mostly in ropes and in water, and the word evokes for us a category of experience. Similarly, we know about particles from balls and pebbles and sand. Light is what it is, but we create experiments that are designed, not to question directly what it is (what would such an experiment be?) but to question whether it can be described in terms of our preexistent categories of thought. Whether the categories are those of particles or waves, the answer is found to be yes. The experiments demonstrate that both categories turn out to be useful to understand light, and this warns us that it is not a wave like water, which cannot possibly be confused with a particle, or a particle like a very small marble, which cannot possibly be confused with a wave. It is, in fact, what it is; its behavior in some experiments can be understood in terms of our previous experience of waves, while others recall particles. Seeking to unify our understanding, we find light described by certain equations of mathematical physics in which we can clearly see that both wave and particle aspects are latent, provided that the appropriate situations are analyzed.

We piece together our understanding of nature from experiments designed using preconceived categories of thought. These categories may fit nature well or badly or not at all. Often, as with waves and particles, two of them cannot be applied at once, and this is a sign that a deeper theory is required to unite them. The theory exists but it is stated in mathematical terms, not in the language of common sensory experience. To explain the theory in ordinary language we must refer to ordinary experience, and experience supplies concepts that are complementary to each other; none is sufficient, but several fill out the picture. The underlying unity appears only at a higher level of abstraction.

I think the complementary nature of Time 1 and Time 2 (or of determinism and freedom) [18] will be understood at a higher level of complexity, in terms of a physical understanding of brain function, but I fear this understanding may be slow in coming and irreducibly complex when we finally have it. [19] One tends to think, following the example of physics, that all the deep questions of science have simple answers, forgetting that physics chooses its problems so that this will be so. Perhaps the reconciliation of Time 1 and Time 2, the explanation of one in terms of the other, is simply not possible in terms that will satisfy the questing mind. But perhaps it is not so important for us today to be able to consider the universe as One. We tend toward pluralism and the belief, embodied in complementarity, that different truths are seen when we look from different angles. Perhaps the only perspective left from which the world can be viewed as a unity is that of physicists and cosmologists, whose imagined world exists only in Time 1. But Time 1 serves another purpose also, for as Socrates said, the unexamined life is not worth living; we cannot be always in the headlong present, swept along in the current of experience. The mode of Parmenides is the one in which we examine ourselves, our conduct, the effect we have on others, and the reasons for our acts. It is true that at that moment the bonds both to the past and the future constrain us and we become less free, but it is then that we take the responsibility for what we do, which makes us Man.

Appendix 1

No Magic in Maxwell's Demon

Since the tendency of the universe to approach equilibrium has been shown to be a statistical matter, and since at first sight statistics seem to be an expression of our ignorance concerning exact causes, it might seem that a hypothetical being who was aware of exact causes could interfere with nature's randomness and violate the second law of thermodynamics. This appendix shows in a simple example how entropy is defined, why it increases, and why an attempt to reverse this increase runs into severe difficulties.

If an amount of heat q is transferred into or out of a region at temperature T degrees above absolute zero, the entropy of the region increases or diminishes by the amount

$$\Delta S = \frac{q}{T}$$

(This is the original thermodynamic definition of entropy.) Suppose for example that the heat q flows from one region (call it the left-hand region) at temperature T_1 into another (right) at T_r. The total entropy change is

$$\Delta S = -\frac{q}{T_1} + \frac{q}{T_r} = q\left(\frac{1}{T_r} - \frac{1}{T_1}\right)$$

As we know, the condition for such a heat flow is that the left-hand region be warmer than the right: $T_1 > T_r$, and in the formula above this makes $\Delta S > 0$, an example of the universal tendency of processes to take place in the direction that increases entropy.

Suppose now that we decide to reverse the flow of heat by hiring a demon (fig. 6) to stand at the door (made of glass) and open it to let in molecules traveling faster than usual from the right side and to let out an equal number of molecules traveling more slowly than usual from the left. The heat flow is reversed from the previous example, and the entropy goes down. Thus it would appear that the demon can violate the second law. But how does the demon know when to open the door? Let us equip him with a flashlight and tell him to use it in the most economical possible way. Being a demon, he is able to adjust it to emit on demand a single quantum of light, which bounces off the molecule into his eye. He has a wonderful eye which responds to color so sensitively that he can tell by the Doppler shift of the light reflected from a molecule whether it is going faster or slower than the average. But he opens the door to only half the molecules he observes, since the other half have the wrong speed. Thus two quanta are used for every molecule that is allowed to pass. Of course, when the demon allows *any* molecule to pass from left to right, he produces an increase in entropy. But suppose he allows a faster-than-average molecule of energy E_r to pass from right to left. The entropy change will be

$$\Delta S = \frac{2E_q}{T_l} + E_r\left(-\frac{1}{T_r} + \frac{1}{T_l}\right)$$

where E_q is the energy of the light quanta he is using. The first term represents the entropy increase as the light quanta are absorbed into the demon's eye, and the other two terms are as above.

The average kinetic energy of particles on the right is $\frac{3}{2}kT_r$, where k is Boltzmann's constant. A simple statistical argument shows that when the demon passes faster-than-average molecules from the right region to the left, their average kinetic energy is $1.81 \times \frac{3}{2} kT_r$. There is an important constraint on the energy E_q, for any region in thermal equilibrium at temperature T is bathed in thermal radiation. The radiation becomes visible when something is made red hot, but it is present at any temperature, and the average of energy of a quantum of this radiation is about $3kT$. For the demon's little flash to be perceptible at all, it must be a quantum more energetic than this: $E_q > 3kT_r$. Putting these results into the last equation gives for the entropy change when this typical molecule is let through

$$\Delta S > \frac{6kT_r}{T_l} - 1.81 \times \frac{3}{2} kT_r\left(-\frac{1}{T_r} + \frac{1}{T_l}\right)$$

or

$$\Delta S > \frac{3}{2} k(1.81 + 2.19 \, \frac{T_r}{T_1}) > 0$$

Thus the demon fails (by a substantial margin) to decrease the world's entropy even though he succeeds in transferring heat from a cooler region to a warmer one. To make headway against the second law he would have to concern himself with suppressing the thermal radiation from the walls of the two chambers and from the gas molecules inside them: he would have to be extremely busy and everywhere at once, and this raises more difficulties than it solves.

Arguments such as this, dealing quantitatively with the entropy cost of information, were the historical origin of modern information theory. The classical paper which began this line of argument is L. Szilard, *Zeitschrift für Physik* 53 (1929): 840. Note that the demon can be exorcised only by one who knows what Maxwell did not know: that energy comes in quanta.

Appendix 2

Spacetime in the Theory of Relativity

A century ago it was known that if a wire is moved near a magnet, or a magnet is moved near a wire, a little voltage will be induced in the wire. This is the principle of the generators that produce our electric power. In a paper published in 1905 [1] Albert Einstein, then an examiner in the Swiss Federal Patent Office, pointed out that although electromagnetic theory accounted perfectly for the phenomena there was something peculiar about the way it did so. Depending on whether it is the wire or the magnet that is moving, two entirely different laws are invoked and the voltage is calculated in entirely different ways. Clearly it should not make any difference, especially since the universe seems to contain no absolute standards of rest and motion. The earth revolves around the sun at 30 kilometers per second, which means that nothing on it can reasonably be said to stand still; yet this has no effect on our ability to play tennis and conduct experiments or any other business of daily life. Thus, although in a certain sense one understood electromagnetic theory, nature seemed to have an element of simplicity, of symmetry, which the theory lacked.

Einstein had been puzzling about various aspects of this disharmony between nature and theory since he was a schoolboy, and it had gradually become clear to him that only a profound modification of the basic principles of physics would allow it to be resolved. [2] He said later that after it occurred to him that "time is the culprit," the whole thing took him only five weeks. This was the genesis of what came to be known as the special theory of relativity. The general theory, published a decade later, incorporates these considerations into a theory of the gravitational field, as has been mentioned in chapter 9.

The phrase "time is the culprit" refers to that part of the theory which provoked the most objections and was the hardest to understand. Let us suppose there is a church tower which we agree to use as a standard for measuring velocities and distances. Then, as we move across the land-scape with a certain velocity of our own, where we are depends on what time it is. This fact is obvious and familiar to everybody. But if we agree to use the clock in the tower as a standard of time Einstein's further con-clusion—that what time it is for us depends on where we are—is *not* obvious. How could the one dependence be a commonplace and the other be entirely unsuspected for millennia? Only because the time effect is, at ordinary rates of travel, very small. There is a fundamental constant built into our description of the fabric of nature, the quantity c mentioned on p. 87, which manifests itself as a limiting speed beyond which no par-ticle can be accelerated in a laboratory. It is, exactly or very nearly, the speed of light. Suppose I drive away from a certain house toward the church at a distance x from it. After I have driven for a time t at velocity v, I have traveled a distance vt and with respect to me the church is at a distance

$$x' = x - vt \tag{1}$$

This is obvious common sense. Common sense further asserts that the clock in the church reads the same time t as the clock in the dashboard of my car, which I shall call t':

$$t' = t \tag{2}$$

It turns out, however, that these two relations are only approximately true. The first must be changed to

$$x' = \gamma(x - vt) \tag{3}$$

where the new scale factor γ depends on my velocity,

$$\gamma = \frac{1}{\sqrt{1 - v^2/c^2}} \tag{4}$$

and the times must be related by

$$t' = \gamma(t - vx/c^2) \tag{5}$$

These relations are known as the Lorentz transformations, after a Dutch physicist who studied these questions before Einstein, though he did not write down the formulas. Einstein's achievement was to derive the formulas, show what they mean, and carry forward the change of ideas implied in them. How the formulas are derived is explained in many texts on elementary physics and two clear accounts by Einstein;[3] it need not be repeated here.

Note that ordinarily the modifications brought by relativity are very small. The speed c is enormous, 2.998×10^5 km/sec, and γ is ordinarily close to 1. Further, for ordinary values of v and x, the extra term in the expression for t' is very small. That is how the need for the corrections was so long unnoticed. Since 1905 we have started building accelerators that speed up particles to velocities very close to that of light, and considerations from Einstein's theory are crucial to their design. If electrons from the beam produced at the Stanford Linear Accelerator raced a beam of light to the moon, the light would win, but only by about 1 cm.

I now point out two algebraic facts about the formulas just given that are not immediately obvious but are easily verified by the ready reckoner with pencil and paper.

1. The formulas give x' and t' in terms of x and t. Solve the two equations for x and t in terms of x' and t'. The result is

$$x = \gamma(x' + vt') \tag{3'}$$

$$t = \gamma(t' + vx'/c^2) \tag{5'}$$

13. A car bearing a clock reading t' travels to a church tower whose clock reads t from a house in a neighboring village.

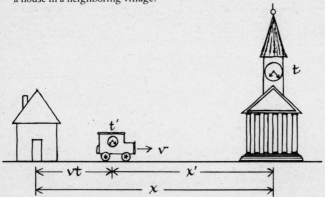

The formulas are of exactly the same shape as the originals except that v is replaced by $-v$. This means simply that if I am moving to the right with respect to you at velocity v, then you are moving to the right with respect to me at velocity $-v$, or, more conventionally, you are moving to the left with respect to me at velocity v. The relative motion is all that counts; in this simple way relativity is contained in the equations.

2. Calculate the quantity $x^2 - c^2t^2$. It comes out to be equal to the corresponding quantity in the other variables:

$$x^2 - c^2t^2 = x'^2 - c^2t'^2 \tag{6}$$

This means that though space intervals and time intervals vary from one observer to another, there is a quantity composed of both together whose value is common to both observers. The theory of relativity has produced an absolute quantity; in fact it has produced two; the other is the speed c. Physicists are people who look for absolutes; perhaps almost the last who do.[4] Equation (6) expresses a symmetry property in space and time taken together, and the union of these two, which is required if the symmetry is to exist, is called spacetime. It contains the three usual dimensions of space: length, breadth, and height, together with a single other dimension, time. It is in this technical sense that time is said to be the fourth dimension.

That space and time are united in this way is forced on us by experiment, but once we understand the relationship expressed in Equation (6), the problem of working out the mathematical structure of spacetime is a purely mathematical one, and in fact the first steps had been taken, as a matter of mathematical interest, before relativity theory appeared. How the mathematical structure is revealed in experience is a question of physical interpretation. That has changed in the past and may change again, but the mathematical structure cannot change (except in being further explored) because it is rooted in proof. People on other worlds may have different sensory apparatus from ourselves and different kinds of physics, but if their mathematical concepts are at all like our own, it seems likely that they will know about mathematical structures like the four-dimensional geometry of spacetime. As I think Heisenberg was the first to point out,[5] these structures, both in their abstract character and in the role they play in our understanding of the world, have a part very similar to that of the Ideas in Plato's theory of knowledge.

Spacetime is absolute in the sense that an abstract mathematical description of it can be made that is the same for all observers. That is, it not only has the formal completeness of mathematical structures but also can be related through definite rules with the results of physical measurements. If we try to subdivide spacetime further into space and time, observers will no longer agree. Space and time are mapped out by measurements also, but observers who share the common spacetime have spaces and times of their own. One observer may judge two spatially separated events as simultaneous, while to another they will be separated by an interval of time. Thus a question like "Why is my time the same as your time?" cannot be answered. They aren't the same, if you and I are walking at different speeds. Only our spacetimes are the same. This is why Einstein and many others have concluded that the space-time description of events, which seems necessarily to be a description in terms of what we have called Time 1, is the fundamental one. It can be expressed in absolute language but not, of course, in terms of the immediate personal experience of the observers we have just been talking about.

The Twins

The strange content of the Lorentz transformations does not become clear until one puts in numbers to see what they say. Let us consider a space voyage, made at a (practically) constant speed of 3/5 the speed of light, to a nearby star 3 light years (ly) away. It will be convenient to measure distance in ly and time in years (y), since the speed of light is 1 ly/y, or just 1. The planners of the mission calculate the time of the voyage to the star as

$$t = \frac{3}{3/5} = 5 \text{ y}$$

and, assuming that the traveler does not like what he sees and starts back at once, he will be gone for 10 years.

But the traveler's watch shows a different time. Substituting the appropriate numbers into (4) and (5), we find for the journey out

$$\gamma = \frac{1}{\sqrt{1-(\frac{3}{5})^2}} = \frac{5}{4}$$

$$t' = \frac{5}{4}(5 - \frac{3}{5} \times 3) = 4y$$

so that the round trip will last, by the traveler's watch, not 10 but 8 years. This result follows from the properties of spacetime and does not refer to the reading of any particular kind of clock. If the space traveler is considered as his own clock, his mind and body will be 8 years older when he returns. In the early days of the theory a number of astonished critics, Henri Bergson among the first, pointed out that if the traveler had left a twin brother behind, they would no longer, by any physical or mental test, be of the same age when they greeted each other after the voyage. For Einstein to have put forward a theory containing a paradox of this kind in the absence of any shred of experimental evidence, merely as part of a program to clarify the laws of electrodynamics, was an act of great courage. In today's physics the unstable particles studied in high-energy laboratories act as little speeding clocks, and Einstein's formula is routinely taken into account in interpreting the results of experiments which, without it, would make no sense.[6]

Appendix 3

Mach's Principle and the Cosmic Numbers

Newton's laws of dynamics (as we write them today) are valid in some frames of reference and not in others. They demand, for example, that the surface of water in a glass will be parallel to the level of the ground; this is not so in a lurching train or on a merry-go-round. The glass must be unaccelerated; that is, not changing its linear velocity and not rotating. But unaccelerated with respect to what? Newton declares: with respect to absolute space. This is an assertion impossible to confirm or deny, since absolute space enters physics only through the laws themselves. (Yet, of course, the world is such that it allows this abstraction to be made and used.) Is there anything more that can be said?

Newtonian dynamics predicts that the orbits of planets will be ellipses that (very nearly) have fixed orientations in absolute space. When telescopic observations are made it is found that the orbits are (to the same accuracy) fixed with respect to the starry background. If one takes literally the terms of Newton's physics this is nothing but a coincidence, for his absolute space has nothing to do with the stars, but in 1872 the Austrian physicist and critic of scientific ideas Ernst Mach (of aerodynamic fame) suggested that the coincidence may be a sign of an unknown but causal connection between Newton's absolute space and the space of the universe.[1]

Mach's idea offered one of the hints that led Einstein to the general theory of relativity, but it is very difficult to state it unambiguously or derive it from the theory.[2] In the 1950s it occurred to several people independently[3] that even if a theoretical argument is lacking, it is possible to guess plausibly what its result would be if there were one. The usual form of Newton's law of acceleration states that

$$f = ma \tag{1}$$

where f is the force required to give an object of mass m an acceleration a with respect to absolute space. This formula gives no hint as to why the force must be exerted. If one assumes that the force is in some way caused by the rest of the universe, then it is natural to implicate gravity, for it acts at long distances, and to guess that the law should be written as

$$f = \frac{GMm}{R^p c^q} a$$

where M is the mass of the universe, c is the speed of light, R has something to do with the size of the universe, and p and q are some unknown exponents that come out of the theory. But we can find p and q without having the theory, for the units must come out right, and this is enough to establish that $p = 1$, $q = 2$, and

$$f = \frac{GMm}{Rc^2} a \tag{2}$$

Comparison with Newton's law (1) gives

$$\frac{GM}{Rc^2} = 1 \tag{3}$$

Now, what do we take for M and R?

We believe that no interaction propagates faster than light. If the universe has been in existence for a time T, then no mass at a distance greater than cT could possibly affect anything that happens here. We express M (leaving out unimportant and unknown constants) as approximately dR^3 where d is the mean density of matter in the universe and replace R by cT. A little algebra then gives

$$GdT^2 \approx 1 \tag{4}$$

We have values for these numbers: in the units of centimeters, grams, and seconds G is about 7×10^{-8}, d is about 3×10^{-31}, and T is about 6×10^{17}. The product is about 0.01; perhaps, in view of our uncertainty over d, that is not too far from 1.

It is hard to say what this means. It is tempting to use it to banish the arbitrary number G from the law of gravity by writing

$$G = \frac{\alpha}{dT^2} \tag{5}$$

where α is some number not very different from 1, so that Newton's law for the gravitational force between masses m_1 and m_2 becomes

$$f_{\text{grav}} = \frac{\alpha}{dT^2} \frac{m_1 m_2}{r^2} \tag{6}$$

a beautiful formula in which properties of the universe as a whole govern the interaction of any two masses. But since d and T both change with time, G defined by (5) is not constant but varies by a few parts in 10^{10} per year. As we have seen in chapter 9, the data probably do not allow this much change, so that even though the relation (4) is found to be roughly true, its meaning remains uncertain.

There are several signs that the general theory of relativity is incomplete: it seems difficult to incorporate gravitational interactions into the quantum theory; the theory does not clearly account for a relation like that deduced here, and it gives nature an absolute and unexplained beginning at time zero as well as an absolute end if it collapses again. There are, of course, deep and subtle speculations going on as to how these difficulties might be surmounted.

Appendix 4

Some Notions of Cosmology

In this appendix I develop some simple dynamical properties of a model universe that expands from a point. Although relativistic considerations of curved space are ultimately involved, the main relations can be well understood through Newtonian reasoning. After developing the Newtonian equation I make the small changes needed to render it relativistic and then comment on a few features of the results.

An Exploding Universe

In the flat space of Newtonian physics there is no way to preserve the idea (so obvious on the surface of a sphere, for example) that all points are equivalent. There must be a point where everything started, where the explosion took place at time zero. In the nonrelativistic theory we must assume further that the explosion had a miraculous character: the velocity distribution of the expanding gas is such that, as seen from the central point, the density of matter at all points and all distances (out to the edge of the expanding universe) shall remain uniform—though of course it decreases all the time. How this happens is shown in figure 14, which illustrates an expanding shell of matter at two instants. As R increases, ΔR also increases proportionately. Thus the two diagrams differ only by a change of scale, and all pairs of points on the second diagram are further apart than they are on the first by the same numerical factor. This means that the universe as seen from O *and from every other point of the expanding gas* will obey Hubble's law that the velocity of recession of surrounding points is proportional to the distance.

From now on we concentrate on a single representative galaxy, α. As was mentioned in chapter 4, it is attracted by the matter in the sphere lying between it and the center, and it can be shown that no mass further out affects it at all. Let its mass be m and that of all the galaxies inside the sphere be M. Then the energy equation, kinetic plus potential equals total energy E, gives

$$\frac{1}{2}\, m \left(\frac{dR}{dt}\right)^2 - G\frac{mM}{R} = E \tag{1}$$

Write M in terms of the smoothed-out matter density d as

$$M = \frac{4}{3}\,\pi R^3 d \tag{2}$$

14. Newtonian cosmology. A spherical shell, centered at an imagined fixed point O and containing a fixed number of galaxies, expands as the galaxies diverge from O. The sphere and the shell are assumed large enough to contain many galaxies, of which a typical one, α, is shown: a, at a given moment; b, at a later moment. Since the expansion is proportional to the distance from O, b is exactly similar to a but on a larger scale.

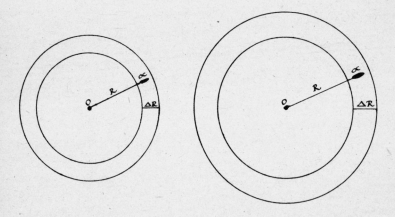

Equation (1) then becomes

$$\left(\frac{dR}{dt}\right)^2 - \frac{2E}{m} = \frac{8\pi}{3}GdR^2 \qquad (3)$$

This simple differential equation describes the motion of any representative point α and therefore determines the behavior of the model as a whole.

What ultimately happens depends on the sign of E. If E is positive, dR/dt is never zero and the system continues to expand. If E is negative, it is clear from (1) that for large enough R the motion stops, and after this a contraction sets in which repeats the expansion in reverse.

Solution of the Equation

If E is negative, (1) represents, in a suitable scale, a cycloid, the path traced out by a point on the rim of a rolling wheel. Let θ be the angle through which the wheel has rolled along the t-axis. It is not hard to see that

$$R = A(1-\cos\theta), \qquad t = B(\theta-\sin\theta) \qquad (4)$$

where A is a scale constant of the dimension of length and B another of the dimension of time. If these are substituted into (1) we find

$$\left[\frac{A\sin\theta}{B(1-\cos\theta)}\right]^2 = \frac{2GM}{A(1-\cos\theta)} = \frac{2E}{m}$$

and this is indeed satisfied for all θ provided that

$$A = -\frac{GmM}{2E}, \qquad B^2 = \frac{A^3}{GM} \qquad (5)$$

R increases from 0 to a maximum of

$$R_{max} = 2A = -\frac{GmM}{E} \qquad (6)$$

(remember that E is negative), while the cycle is complete when $\theta = 2\pi$ and the duration of a cycle is

$$T_{cyc} = 2\pi B = 2\pi\sqrt{(R_{max}^3/8GM)}$$

Remembering (2), we can express this in terms of d_{min}, the spatial density of matter when R has its largest value:

$$M = \frac{4\pi}{3} R_{max}{}^3 d_{min}$$

and in terms of $d_{min'}$

$$T_{cyc} = \sqrt{(3\pi/8Gd_{min})} \tag{7}$$

A similar integration can be performed if E is positive or zero, and the results are shown qualitatively in figure 15. All these results, however, are formal and artificial, belonging to a Newtonian model that happens to expand at uniform density. It is remarkable that they can be carried over almost without change into a believable relativistic theory.

Relations in Curved Space

It will be clearest to start with the equation of motion in the form (3), and let R now denote, instead of the radius from an assumed center to an arbitrary nebula α, the radius of curvature of the space itself. If R is

15. Evolution of R corresponding to positive and negative values of E. The slope dR/dt of the line tangent to the curve at the present moment, t_p, gives Hubble's constant, H. T is the overestimate of the age of the universe which one obtains simply from H.

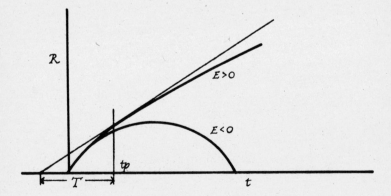

positive, space is curved so as to produce a finite total volume, uniformly filled with matter, exactly as a two-dimensional surface can be curved into the surface of a sphere. If R is negative, the space is infinite in extent and has a curvature which in three dimensions is hard to visualize; it corresponds to the doubly-curved shape of a Western saddle in two dimensions. If R is infinite the space is flat. The general theory of relativity links the properties of space to its material contents (chap. 9), and when the field equations of the theory are written down for a simple choice of coordinates in this simply structured space it is found that E in (3) is replaced by $-\frac{1}{2}\,mc^2$, $+\frac{1}{2}\,mc^2$, or 0 corresponding to the three possible cases mentioned earlier. There is one other change. Pressure produces energy, and energy, in relativity theory, has mass. Counter to what one might at first think, the effect of pressure in gas or radiation is to retard expansion and hasten collapse: the pressure does not act outward because there is no outward; all directions are equivalent. All it does is to produce energy that increases the effective value of d. In our universe however, the pressure is very small. It can only be important near the beginning of the motion, and at the end if the universe collapses again.

Neglecting the pressure, we can take over all the earlier results. Equation (6) becomes (for a closed universe)

$$R_{\max} = \frac{2GM}{c^2} \qquad (8)$$

while (7) remains unchanged.

The Fate of the Universe

None of the foregoing relations can be directly verified because we have no way of knowing the value of R or of $d_{\min}$. If we believe the theory, however, we can use it to decide whether the universe is destined to go on expanding forever or whether it will ultimately collapse. We have seen that this depends only on the sign of the quantity represented by E in (1): if E is positive, expansion continues. This evidently requires

$$\frac{1}{2}\,m\left(\frac{dR}{dt}\right)^2 > \frac{GmM}{R}$$

or, in terms of the density d,

$$\frac{8\pi}{3}\,Gd\,\frac{R^2}{(dR/dt)^2} < 1$$

Consider the motion of galaxies at a distance λR from the observer, where λ is some small fraction. The velocity of the motion is $\lambda dR/dt$, and by Hubble's law this is proportional to the distance:

$$\lambda \frac{dR}{dt} = H\lambda R$$

Thus

$$\frac{dR}{dt} = HR \qquad (9)$$

where H, called Hubble's constant, is the reciprocal of the quantity T in chapter 9 which approximates the age of the universe,

$$H = 1/T \qquad (10)$$

The criterion for continued expansion is

$$\Omega < 1, \text{ where } \Omega = \frac{8\pi}{3} \frac{Gd}{H^2} \qquad (11)$$

and all the Latin letters denote quantities that can in principle be measured. Numerical values for G and d were given at the end of Appendix 2. The proposed values of Hubble's constant have varied widely over the years, but a reasonable modern value is

$$H = 55 \frac{km/sec}{megaparsec} = 1.8 \times 10^{-18} \ sec^{-1}$$

With these values, the quantity on the left of (11) comes out to be 0.05. Apparently the universe will continue to expand forever, unless an error of a factor of about 20 can be found in the data or the approximate theory.

Perhaps it is a curious fact that there is any doubt at all as to whether (11) is satisfied. Most numbers in astronomy are very far from 1. If it is not a coincidence that Ω is in the neighborhood of 1, it is perhaps explainable by some future theory of the universe such as that hinted at by the arguments of Appendix 2. There Equation (4), expressed in terms of H, reads

$$\frac{Gd}{H^2} \approx 1$$

This places Ω near 1, and it perhaps explains why it is so hard to find out whether it is a little greater or a little smaller.

If all other explanations fail, there is still an easy way to understand why Ω is about 1. Write

$$\frac{1}{\Omega} = \frac{3}{8\pi} \frac{H^2}{Gd}$$

and use (2) and (9) to get

$$\frac{1}{\Omega} = \frac{R}{2MG} \left(\frac{dR}{dt}\right)^2$$

Now substitute (1) for $(dR/dt)^2$ and use (2) again:

$$\frac{1}{\Omega} = 1 + \frac{E}{mMG} R$$

or

$$\Omega = \frac{1}{1 + \dfrac{E}{mMG} R} \tag{12}$$

At the beginning of the universe, $R = 0$ and $\Omega = 1$. As R rapidly increases thereafter, Ω increases if $E < 0$ (the case considered in detail above) and decreases if $E > 0$. If $E < 0$, it follows from (6) that Ω becomes infinite when R reaches its maximum, while if $E > 0$ it just gets smaller and smaller. Thus if Ω seems to be close to 1, observation may simply be telling us that the universe is still young, a fact we might have guessed in any case from the many signs of its first creation that lie all about us.[1]

Notes

Chapter 1

1. Book 11 of the *Confessions*.
2. The relation of time to space belongs to the theory of relativity and is discussed in Appendix 2.
3. A simple proof in the Greek spirit is given by Bronowski 1973. For Babylonian mathematics see Neugebauer 1952, and van der Waerden 1954.
4. Several essays on this subject are collected in Gale 1967.
5. *On Interpretation*, 18b, my adaptation from the convenient edition Aristotle 1941; see also Gale 1967.

Chapter 2

1. See Hoyle 1977. Two other accounts of Stonehenge, though based on careful measurement and calculation, may impress some readers as occasionally over-speculative: Hawkins 1965, and A. Thom, A. S. Thom, and A. S. Thom, "Stonehenge," *Journal of the History of Astronomy* 5(1974): 71.
2. E. C. Baity, "Archaeoastronomy and Ethnoastronomy So Far," *Current Anthropology* 14(1973): 389, gives a survey with more than 600 references, followed by critical comments and more references by a number of specialists. See also Brown 1976 and Wood 1978.
3. See for example Eliade 1954.

Chapter 3

1. For a notable and well-informed attempt see Lockyer 1894. An extremely interesting and imaginative work which however draws conclusions far beyond its evidence is de Santillana and von Dechend 1969. See also Hartner 1968.
2. E. Zilsel, "The Genesis of the Concept of Physical Law," *Philosophical Review* 51(1942): 245; F. Oakley, "Christian Theology and Newtonian Science: The Rise

of the Concept of Laws of Nature," *Church History* 30(1961): 343, reprinted in O'Connor and Oakley 1969.

3. *Physics*, 203b.

4. Anaximander's law is that of the balance beam, which leans one way or the other according to the weights on it but always comes to rest in equilibrium. This is the fundamental motion of Greek drama and, after much intervening history, of our own. A perfect example is Sophocles' *Oedipus at Colonus*: not only is the play constructed thus but in it Oedipus refers twice to the same motion in describing his own life. In Christian representations of the Last Judgment, Christ is shown weighing souls. We take this metaphorically, but the fusion of physical and moral law has a strong feeling of antiquity.

5. A brilliant essay, "The Scar of Achilles" in Erich Auerbach's *Mimesis* (1953), contrasts this "presentness" of Homer with the narrative technique of the Old Testament in which detail is suppressed and the importance of the story is seen in its overall significance.

6. Simplicius as quoted in Kirk and Raven 1957, p. 415.

7. Theophrastus as quoted in Kirk and Raven 1957, p. 423.

8. Simplicius as quoted in Kirk and Raven 1957, p. 419.

9. The most complete statement of atomic theory we have from antiquity is that of Lucretius, whose *De Rerum Natura* expounds a natural philosophy based on atomism. At about the same time as Lucretius, atomistic philosophy arose in India, though whether it was indigenous or traveled across the trade routes of Bactria is not certain. At any rate, Indian atoms are very different from Greek, since they are understood in terms of the sensations they produce rather than the matter they constitute.

Chapter 4

1. See, for example, de Santillana and von Dechend 1969.

2. The distinction between science and nonscience has been argued with special force and clarity by Karl Popper. See the first essay, "Science: Conjectures and Refutations" in Popper 1962; see also Popper 1959.

3. Having no object isolated from all external forces to study, Galileo derived his conclusion from indirect arguments, but it was at least empirically based (Galileo 1914, p. 244). Aristotle had reached it long before on logical grounds. Talking about motion in a vacuum where there would be no force exerted by the air, he said, "No one can say why a thing once set in motion should stop at any particular point: for why should it stop here rather than there? Thus a thing will either remain at rest or move forever unless something more powerful gets in its way" (*Physics*, 215a, adapted from Aristotle 1941). Aristotle found this conclusion manifestly absurd, and he used it to argue that there can be no vacuum anywhere.

4. The question of why Newton allowed speculation to go on for 20 years more when he believed he knew the answer has been much discussed. One explanation is that the numbers did not come out right until a new survey corrected by a few percent the then-current figure for the diameter of the earth. Another is that he had at first no way of estimating the effect of the sun's attraction. A third, which can be supported by modern examples advanced in an interesting recent book (Dirac

1971), is that he was scared. A scientist who has done something new may be so overwhelmed by the consequences that it will have if right, and so afraid it might be wrong, that he is for a while unable to explore it further. At any rate, Newton had other scientific projects to occupy him.

5. In particular, Newton's derivation of the precession of the equinoxes was very primitive, a work of genius precisely because in the absence of any dynamical theory he had to base his whole derivation on clever analogies. It is in fact not really a derivation at all, though he tried to make it appear so. See R. S. Westfall, "Newton and the Fudge Factor," *Science* 179(1973): 751.

6. See especially Jammer 1969. The literature on Newton is immense; a small but vital part of it is Koyré 1965.

7. Minkowski in Lorentz *et al*. 1923.

Chapter 5

1. J. V. Noble and D. J. de Solla Price, "The Water Clock in the Tower of the Winds," *American Journal of Archaeology* 72(1968): 345; D. J. de Solla Price in Fraser and Lawrence 1975.

2. Note that Uranos has the same name as Varuna, the Indic sky-god.

3. Price in Fraser and Lawrence 1975.

4. For an interesting survey of the principles and the intellectual problems of time measurement, see Fraser 1975, pp. 47–63.

5. This argument was first given in 1754 by Immanuel Kant 1900. A numerical estimate of the effect is difficult and, not surprisingly, Kant's is far from accurate. It has been proposed that there is another reason why the celestial clocks disagree: the Newtonian gravitational constant G is not a constant but has slowly decreased through the millennia. This would affect the orbital periods of earth and moon. It would also cause the earth to expand slightly, slowing its rotation. There is evidence both for and against this proposal, but at present that against it appears much stronger. It is discussed further in chapter 9.

Chapter 6

1. Very roughly, one million dollars is a pile of $1,000 bills 6 inches high. One billion dollars is a pile as high as the Washington Monument. To visualize the defense budget one need imagine the pile laid on its side along a road. At about $10 billion per mile, it would extend some 10 miles.

2. This is the small amount of heat required to raise 1 gram of water 1 degree centigrade. It is one-thousandth of the "calorie" known to dieters, which is more properly called a kilocalorie.

3. L. Szilard, "Über die Entropieverminderung in einem thermodynamischen System bei Eingriffen Intelligenter Wesen," *Zeitschrift für Physik* 53(1929): 840; see also L. Brillouin, "Maxwell's Demon Cannot Operate: Information and Entropy. I," *Journal of Applied Physics* 22(1951): 334.

4. P. T. Landsberg, "Thermodynamics, Cosmology, and the Physical Constants," in Fraser, Lawrence, and Park 1978, argues the contrary.

Chapter 7

1. The problems of trying to go further than this are seen in old attempts to explain why the planets obey the laws governing their motion. Plato saw the motions of matter governed by principles of reason and beauty, which the planets obey as "living beings, divine and everlasting." This mythical language derived from an ancient system of ideas, but Johannes Kepler was forced to essentially the same conclusion, and he did not think of himself as a mythographer.

Chapter 8

1. A. Burl, "Dating the British Stone Circles," *American Scientist*, Mar.–Apr. 1973, p. 167.
2. In the first account of creation in Genesis, the authors show their independence of the mythological current by specifying that God created the watery chaos also. The "spirit of God" moving over the water means at the same time breath, wind, or storm of God, and the famous firmament, *rāqia'*, is an artifact, a bell-shaped vault of hammered metal which separates the waters from the waters. See von Rad 1972.
3. *On the Heavens*, 298a.
4. *Politics*, 1265a.

Chapter 9

1. An idea of how it might have been can be obtained from Marcel Griaule's (1965) sensitive account of his initiation into the cosmology of the Dogon tribe of West Africa. This cosmology has some remarkable affinities with that of Mesopotamia and the Mediterranean.
2. An interesting discussion of curved space considered intrinsically, without the aid of extra dimensions, is given by J. J. Callahan, "The Curvature of Space in a Finite Universe," *Scientific American* 235(August 1976): 90.
3. Since the theory of relativity is based on experimental observations, it is hard to learn to understand it without submitting in some degree to the discipline of the universe. There are many books that try to help. Einstein 1920, Einstein and Infeld 1938, Bondi 1965, Bergmann 1968, and Davies 1977 are worth trying.
4. There is a very good book about all this: Weinberg 1977.
5. The law of electric force is valid over 34 orders of magnitude in distance. The same law that governs the number of neutrons and protons in a stable nucleus also governs the stability of a neutron star, a range in mass of 55 orders of magnitude. It was Newton who first began to grasp the range of validity of quantitative physical laws. Perhaps it was his greatest discovery.
6. P. A. M. Dirac, "The Cosmological Constants," *Nature* 139(1937): 323. Some of the continuing discussion can be found in Dirac's contribution to Mehra 1973, Dirac and Teller in Reines 1972, and Dyson in Salam and Wigner 1972.
7. T. C. Van Flandern in Bergmann, Fenyves, and Motz 1975. See also "A Determination of the Rate of Change of G," *Monthly Notices, Royal Astronomical Society* 170(1975): 333; F. Hoyle, "The History of the Earth," *Quarterly Journal Royal*

Astronomical Society 13(1972): 328, suggests a decrease of about 6 percent per 10^9 years.

8.   R. H. Dicke, "Dirac's Cosmology and Mach's Principle," *Nature* 192(1961): 440.

9.   P. A. M. Dirac, "Reply to Dicke," *Nature* 192(1961): 441.

10.   A. M. Wolfe, R. L. Brown, and M. S. Roberts, "Limits on the Variation of Fundamental Atomic Quantities over Cosmic Time Scales," *Physical Review Letters* 37(1976): 179. There are less direct arguments that put much sharper limits on this and other atomic constants: See A. L. Shlyakhter, "Direct Test of the Constancy of Fundamental Nuclear Constants," *Nature* 264(1976): 340.

11.   In Longair 1974. There is a growing number of instances in which it appears that the sizes of the numbers determining the properties of the universe are those which permit some form of life to emerge. See B. J. Carr and M. J. Rees, "The Anthropic Principle and the Structure of the Physical World," *Nature* 278(1979): 605 and Dyson 1979, chapter 23.

12.   S. W. Hawking, "Breakdown of Predictability in Gravitational Collapse," *Physical Review* D14(1976): 2460. The ample and finely written introduction to this paper will instruct even those who do not follow mathematical derivations.

13.   For further discussion see Davies 1974, 1977, and his essay in Fraser, Park, and Lawrence 1977; also D. Layzer, "The Arrow of Time," *Scientific American* 236 (December 1975): 56.

14.   P. T. Landsberg and D. Park, "Entropy in an Oscillating Universe," *Proceedings of Royal Society of London* A346(1975): 485.

15.   See for example Augustine, *Confessions*, Bk. 11.

16.   Popper 1962, p. 28.

Chapter 10

1.   This should be said more carefully. Judged by their descriptions of celestial motions, geocentric and heliocentric systems are exactly equivalent, but that is not the only criterion; there are two others. The first is provided by the dynamical laws that account for the motions. If we use the geocentric system they are not only complicated and difficult to apply, but they also contain quantities referring to the earth's motion and so could not claim universal status: it would be hard to imagine astronomers in a distant galaxy using dynamical laws that referred explicitly to our earth. In the heliocentric system the laws are not only simple but truly universal. The other criterion is one of meaning. The Copernican system leads to a view of man in the universe that finally seems more just than that which underlies the Ptolemaic and scriptural one. The two criteria are not independent of one another.

2.   Bergson 1910, p. 172.

3.   Skinner 1971, e.g., p. 21.

4.   Cornford 1937 has given a thoroughly annotated edition of the *Timaeus*. The translation quoted here is that of Lee 1965; the passages quoted are from pp. 50 and 51 in Lee, corresponding to pp. 94, 97 in Cornford or, in the usual notation, *Timaeus* 37C, 38C.

5.   Plato, *Parmenides*, 138C, 141E.

6.   *Timaeus*, 38C; Lee 1965, p. 51.

7. The standard modern-language edition is that of Festugière: Proclus 1968, but there is a good English translation by the English Platonist Thomas Taylor 1820. The translations given here are somewhat condensed English versions of Festugière.

8. Proclus 1968, 4: 73, or Taylor 1820, 2: 217.

9. Actually, Newton never saw this equation, which was only written down by Leonhard Euler after his death. The mathematical methods used in the *Principia* are cumbersome and obsolete; the only equations relate to the properties of geometric figures. William Whewell wrote of Newton's mathematics in the *Principia*, "The ponderous instrument of synthesis, so effective in his hands, has never since been grasped by one who could use it for such purposes; and we gaze at it with admiring curiosity, as on some gigantic implement of war, which stands idle among the memorials of ancient days, and makes us wonder what manner of man he was who could wield as a weapon what we can hardly lift as a burden." (Whewell 1869, vol. 1, p. 408). Nevertheless, I shall refer to the equation as Newton's.

10. For *cognoscenti*, I refer to Fock's analysis of the hydrogen atom in terms of the four-dimensional orthogonal group $O(4)$ and the role of the special unitary group $SU(3)$ in the dynamics of general three-dimensional systems as well as in the transformations of elementary particles.

11. Proclus 1968, 4: 49, or Taylor 1820, 2: 201.

12. Bergmann 1929.

13. The classic work in this field has been done by Jean Piaget. See for example Piaget 1953; Piaget and Inhelder 1956.

14. It helps to read Körner 1955 first.

15. Bergson 1910, p. 100; see also Herrigel 1953.

16. A number of authors have explained how to measure the changing world against this scale. See, for example, D. C. Williams, "The Myth of Passage" in Williams 1966; chapter 7 of Schlick 1949; and J. J. C. Smart, "Spatializing Time," *Mind* 64(1965): 239. These are all reprinted in Smart 1964.

17. The reader who wishes to dig deeply must go to Bohr, but be warned that he needs to be read with a microscope. It is well to start with Bohr 1963. See also Heisenberg 1938 and, to relate these ideas to what has been thought before, Petersen 1968.

18. Brief but cogent remarks on this subject are made by Bohr 1934, pp. 110, 116.

19. The philosophical territory on which relations of this kind are to be established has been mapped by J. T. Fraser 1975, see also 1978, in a work of great courage and learning in which he takes the relation of times discussed here as the prototype of others that finally form a hierarchy of orders in the universe connecting levels of matter and organization as we understand them. The problem of unification I have discussed does not correspond exactly with any of those in Fraser's books, but the problems are similar even if the ideas of what constitutes an answer are different.

Appendix 2

1. Translated in Lorentz, Einstein, Minkowski, and Weyl 1923. The original paper is Einstein, "Zur Elektrodynamik bewegter Körper," *Annalen der Physik* 17 (1905): 891.

2.   For Einstein's autobiography see Schilpp 1949 or Einstein 1979.

3.   Einstein 1920; Einstein and Infeld 1938.

4.   The theory only gradually acquired the name of relativity. It was not Einstein's fault.

5.   Heisenberg 1938, p. 59. See also Heisenberg 1974, especially the essay "Natural Law and the Structure of Matter."

6.   The question raised here is well discussed in Marder 1971. See also Terletskii 1968 and, for a good introduction to the theory and its modern applications, Taylor and Wheeler 1963.

Appendix 3

1.   Mach 1872; Mach 1883, chapter 2, sec. 6.

2.   The differential geometry necessary to solve problems like this has made great progress in recent years, and a paper by D. J. Raine, "Mach's Principle in General Relativity," *Monthly Notices, Royal Astronomical Society* 171(1975): 507, establishes results that should be noted here. "Einstein gave the name 'Mach's Principle' to the following related ideas: that only relative motion is observable, and hence that there should be no dynamically privileged reference frames; that inertial forces should arise from a gravitational interaction between matter only, and so from an observer-dependent splitting of the total gravitational field; that spacetime is not an absolute element of physics, but that its metric structure is totally dependent on the matter content of the Universe." Raine shows that if this principle is adopted as a postulate, many cosmological models allowed by general relativity are forbidden: the universe cannot be empty of matter; if it contains a uniform distribution of matter and is expanding, the expansion must be the same in every direction; the universe cannot be in rotation as a whole. All these conditions are satisfied by the universe we observe, and we are able to say that Mach's principle is in agreement with everything we know. One could not claim to have established its truth unless one could show that no "non-Machian" model can account for the observations, and that seems very unlikely.

3.   D. W. Sciama, "Origin of Inertia," *Monthly Notices, Royal Astronomical Society* 113 (1953): 34, Sciama 1959; D. Park, "La Notion de particule en théorie classique et en théorie quantique," *Journal de Physique et le radium* 18(1957): 11; R. H. Dicke, "Gravitation—An Enigma," *American Scientist* 47(1959): 25; "Dirac's Cosmology and Mach's Principle," *Nature* 192(1961): 440.

Appendix 4

1.   For more extended discussions without formulas see Laurie 1974; Motz 1975; Goldsmith 1976; and Shipman 1976.

Bibliography

List A contains readable and interesting books directed at the general reader that I used, or could have used, in preparing this one. List B contains books of a more (but not very) specialized character. I have included as many paperbacks as possible. In general, the date after an author's name corresponds to the first edition (or the first edition of a translation). The date of the paperback, where available, corresponds to the latest paperback edition.

List A

Aristotle 1941. *The Basic Works*, ed. R. McKeon. New York: Random House.

Auerbach, E. 1953. *Mimesis: The Representation of Reality in Western Literature*. Princeton: Princeton Univ. Press.

Bergson, H. 1910. *Time and Free Will*. New York: Macmillan; pap. Harper-Row.

Bohr, N. 1934. *Atomic Theory and the Description of Nature*. Cambridge: Cambridge Univ. Press; pap. Cambridge, 1961.

————. 1963. *Essays 1958–1962 on Atomic Physics and Human Knowledge*. New York: Wiley; pap. Vintage, 1966.

Bondi, H. 1965. *Relativity and Common Sense*. Garden City, N.Y.: Anchor Books.

Born, M., ed. 1971. *The Born-Einstein Letters*. New York: Walker.

Bronowski, J. 1973. *The Ascent of Man*. Boston: Little, Brown.

Brown, P. L. 1976. *Megaliths, Myths, and Men*. Poole: Blandford Press.

Burl, A. 1975. *The Stone Circles of the British Isles*. New Haven: Yale Univ. Press.

Burtt, E. A. 1924. *The Metaphysical Foundations of Modern Physical Science*. New York: Harcourt, Brace; pap. Doubleday, 1954.

Calder, N. 1970. *The Violent Universe*. New York: Viking.

Coe, M. D. 1971. *The Maya*. London: Penguin Books.

Cornford, F. M. 1937. *Plato's Cosmology*. London: Routledge & Kegan Paul; pap. Library of Liberal Arts (Bobbs-Merrill).

————. 1939. *Plato and Parmenides*. London: Routledge & Kegan Paul; pap. Liberal Arts Press, 1957.

Davies, P. C. W. 1977. *Space and Time in the Modern Universe*. Cambridge: Cambridge Univ. Press.

de Santillana, G. 1961. *The Origins of Scientific Thought*. Chicago: University of Chicago Press.

de Santillana, G. and von Dechend, H. 1969. *Hamlet's Mill*. Boston: Gambit. This book on the astronomical origins of myth is sufficiently controversial that the careful reader ought also to consult two reviews, by a historian and an astronomer respectively: L. White, Jr. *Isis* 61(1970): 540; C. Payne-Gaposchkin. *Journal for the History of Astronomy* 3(1972): 206.

Dirac, P. A. M. 1971. *The Development of Quantum Theory*. New York: Gordon & Breach.

Dyson, F. 1979. *Disturbing the Universe*. New York: Harper and Row.

Einstein, A. 1920. *Relativity, the Special and the General Theory*. London: Methuen; pap. Crown, 1961.

―――. 1979. *Autobiographical Notes*. Chicago: Open Court.

Einstein, A. and Infeld, L. 1938. *The Evolution of Physics*. New York: Simon & Schuster; pap. Simon & Schuster, 1961.

Eliade, M. 1954. *The Myth of the Eternal Return*. New York: Pantheon. Reprinted as *Cosmos and History*. New York: Harper, 1959.

Feynman, R. P. 1965. *The Character of Physical Law*. London: B.B.C.; Cambridge, Mass.: M.I.T. Press, 1965; pap. M.I.T. Press, 1967.

Fraser, J. T., ed. 1966. *The Voices of Time*. New York: Braziller.

―――. 1975. *Of Time, Passion, and Knowledge*. New York: Braziller.

―――. 1978. *Time as Conflict*. Basel: Birkhaüser.

Fraser, J. T. and Lawrence, N., eds. 1975. *The Study of Time II*. New York: Springer-Verlag.

Fraser, J. T., Lawrence, N., and Park, D., eds. 1978. *The Study of Time III*. New York: Springer-Verlag.

Freeman, K. 1948. *Ancilla to the Pre-Socratic Philosophers*. Cambridge, Mass.: Harvard Univ. Press.

―――. 1959. *Companion to the Pre-Socratic Philosophers*. Cambridge, Mass.: Harvard Univ. Press.

Gale, R. M., ed. 1967. *The Philosophy of Time*. New York: Doubleday; pap. Macmillan 1968.

Galileo, G. 1914. *Dialogues Concerning Two New Sciences*. Translated by H. Crew and A. de Salvio. New York: Macmillan; pap. Dover.

Gold, T. and Schumacher, D. I., eds. 1967. *The Nature of Time*. Ithaca, N.Y.: Cornell Univ. Press.

Goldsmith, D. 1976. *The Universe*. Menlo Park, Calif.: W. A. Benjamin and Co.

Griaule, M. 1965. *Conversations with Ogotemmeli*. Oxford: Clarendon Press; pap. Oxford.

Harnard, S. R., Steklis, H. D., and Lancaster, J., eds. 1977. *Origins and Evolution of Language and Speech*. New York: New York Academy of Sciences.

Hartner, W. 1968. *Oriens, Occidens*. Hildesheim: Olms.

Hawkins, G. 1965. *Stonehenge Decoded*. New York: Doubleday; pap. Delta Books, 1966.

Heidel, A. 1951. *The Babylonian Genesis*. Chicago: University of Chicago Press; pap. University of Chicago.

Heisenberg, W. 1938. *Physics and Philosophy*. New York: Harper; pap. Harper, 1962.

―――. 1974. *Across the Frontiers*. New York: Harper & Row.

Herrigel, E. 1953. *Zen in the Art of Archery*. New York: Pantheon Books; pap. Vintage, 1971.

Hoyle, F. 1977. *On Stonehenge*. San Francisco: Freeman; pap. 1977.

Jammer, M. 1969. *Concepts of Space*. 2d ed. Cambridge, Mass.: Harvard Univ. Press; pap. Harvard, 1969.

Kant, I. 1900. *Cosmogony*. Translated by W. Hastie. Glasgow: Maclehose. Reprinted, with a historical introduction by G. J. Whitrow. New York: Johnson, 1970.

Kirk, G. S. and Raven, J. E. 1957. *The Presocratic Philosophers*. Cambridge: Cambridge Univ. Press; pap. Cambridge, 1969.

Körner, S. 1955. *Kant*. London: Penguin Books.

Koyré, A. 1965. *Newtonian Studies*. Cambridge, Mass.: Harvard Univ. Press; pap. University of Chicago Press, 1968.

Laurie, J., ed. 1974. *Cosmology Now*. London: B.B.C.

Lockyer, J. N. 1894. *The Dawn of Astronomy*. New York: Macmillan; pap. M.I.T. Press, 1964.

Mach, E. 1872. *The Conservation of Energy*. Chicago: Open Court.

————. 1883. *The Science of Mechanics*. Chicago: Open Court.

Marder, L. 1971. *Time and the Space-Traveller*. Philadelphia: University of Pennsylvania Press.

Marshack, A. 1972. *The Roots of Civilization*. New York: McGraw-Hill.

Mendelssohn, K. 1974. *The Riddle of the Pyramids*. New York: Praeger.

Merleau-Ponty, J. and Morando, B. 1976. *The Rebirth of Cosmology*. New York: Knopf.

Motz, L. 1975. *The Universe: Its Beginning and End*. New York: Scribner.

Neugebauer, O. 1952. *The Exact Sciences in Antiquity*. Princeton: Princeton Univ. Press; pap. Harper, 1962.

O'Connor, D. and Oakley, F., eds. 1969. *Creation: The Impact of an Idea*. New York: Scribner.

Petersen, A. 1968. *Quantum Physics and the Philosophical Tradition*. Cambridge, Mass.: M.I.T. Press.

Piaget, J. 1969. *The Child's Conception of Time*. Translated by A. J. Pomerans. London: Routledge & Kegan Paul; pap. Ballantine, 1971.

————. 1970. *The Child's Conception of Movement and Speed*. Translated by G. E. T. Holloway and M. J. Mackenzie. London: Routledge & Kegan Paul; pap. Ballantine, 1971.

Piaget, J. and Inhelder, B. 1956. *The Child's Conception of Space*. London: Routledge & Kegan Paul; pap. Norton, 1967.

Plato 1971. *Timaeus and Critias*. Translated by D. Lee. London: Penguin Books.

Popper, K. R. 1959. *The Logic of Scientific Discovery*. New York: Basic Books; pap. Harper, 1968.

————. 1962. *Conjectures and Refutations*. New York: Basic Books; pap. Harper, 1968.

Proclus 1968. *Commentaire sur le Timée*. Translated by A. J. Festugière. Paris: Vrin.

Reines, F., ed. 1972. *Cosmology, Fusion, and Other Matters*. Boulder: Colorado Assoc. Univ. Press.

Rosenberg, G. D. and Runcorn, S. K., eds. 1975. *Growth Rhythms and the History of the Earth's Rotation*. New York: Wiley.

Schilpp, P. A., ed. 1949. *Albert Einstein, Philosopher-Scientist*. Evanston: Library of Living Philosophers; pap. Open Court, 1973.

Schlick, M. 1949. *The Philosophy of Nature*. Translated by A. von Zeppelin. Philadelphia: Philosophical Library.

Sciama, D. W. 1959. *The Unity of the Universe*. Garden City, N.Y.: Doubleday; pap. Anchor Books, 1961.

Shipman, H. L. 1976. *Black Holes, Quasars, and the Universe*. Boston: Houghton Mifflin; pap. Houghton Mifflin, 1976.

Skinner, B. F. 1971. *Beyond Freedom and Dignity*. New York: Knopf.

Smart, J. J. C., ed. 1964. *Problems of Space and Time*. New York: Macmillan; pap. Macmillan, 1964.

Taylor, T. 1820. *The Commentaries of Proclus on the Timaeus of Plato*. London: Valpy.

Terletskii, Ya. P. 1968. *Paradoxes in the Theory of Relativity*. New York: Plenum Pub.

Thom, A. 1967. *Megalithic Sites in Britain*. Oxford: Clarendon Press.

———. 1971. *Megalithic Lunar Observatories*. Oxford: Clarendon Press.

———. 1979. *Megalithic Remains in Britain and Brittany*. Oxford: Clarendon Press.

Toulmin, S. and Goodfield, J. 1965. *The Discovery of Time*. New York: Harper; pap. Harper, 1966.

van der Waerden, B. L. 1954. *Science Awakening*. Gronigen: Noordhoff; pap. Wiley, 1963.

Von Rad, G. 1972. *Genesis*. Philadelphia: Westminster Press.

Weinberg, S. 1977. *The First Three Minutes: A Modern View of the Origin of the Universe*. New York: Basic Books; pap. Bantam, 1979.

Whewell, W. 1869. *History of the Inductive Sciences from the Earliest to the Present Time*, 3rd ed. New York: Appleton.

Whitehead, A. N. 1925. *Science and the Modern World*. New York: Macmillan; pap. Free Press, 1967.

Whitrow, G. J. 1961. *The Natural Philosophy of Time*. London: Nelson; pap. Harper.

———. 1972. *The Nature of Time*. New York: Holt, Rinehart, and Winston. Published in England by Thames and Hudson, 1972 as *What is Time?*

Williams, D. C. 1966. *Principles of Empirical Realism*. Springfield, Ill.: Thomas.

Wood, J. E. 1978. *Sun, Moon, and Standing Stones*. Oxford: Oxford Univ. Press.

Wright, T. 1750. *An Original Theory or New Hypothesis of the Universe*. London: Chapelle; Reprinted by MacDonald-Elsevier, 1971.

List B

Bergmann, H. 1929. *Der Kampf um das Kausalgesetz in der jüngsten Physik*. Braunschweig: Vieweg.

Bergmann, P. G. 1968. *The Riddle of Gravitation*. New York: Scribner.

Bergmann, P. G., Fenyves, E. J., and Motz, L., eds. 1975. *Seventh Texas Symposium on Relativistic Astrophysics*. New York: New York Academy of Sciences.

Berry, M. 1976. *Principles of Cosmology and Gravitation*. Cambridge: Cambridge Univ. Press.

Bondi, H. 1969. *Cosmology*. 2d ed. Cambridge: Cambridge Univ. Press.

Davies, P. C. W. 1974. *The Physics of Time Asymmetry*. Berkeley: University of California Press.

Fowler, W. A. 1967. *Nuclear Astrophysics*. Philadelphia: American Philosophical Society.

Longair, M. S., ed. 1974. *Confrontation of Cosmological Theories with Observation*. Dordrecht: Reidel.

Lorentz, H. A., Einstein, A., Minkowski, H., and Weyl, H. 1923. *The Principle of Relativity*. New York: Dodd, Mead; pap. Dover.

Mehra, J., ed. 1973. *The Physicist's Conception of Nature*. Dordrecht: Reidel.

Peebles, J. E. 1971. *Physical Cosmology*. Princeton: Princeton Univ. Press; pap. Princeton, 1971.

Rees, M., Ruffini, R., and Wheeler, J. A. 1974. *Black Holes, Gravitational Waves, and Cosmology*. New York: Gordon & Breach.

Salam, A. and Wigner, E. P., eds. 1972. *Aspects of the Quantum Theory*. Cambridge: Cambridge Univ. Press.

Taylor, E. F. and Wheeler, J. A. 1963. *Spacetime Physics*. San Francisco: Freeman.

Index